List of titles

Already published

Cell Differentiation	J. M. Ashworth
Biochemical Genetics	R. A. Woods
Functions of Biological Membranes	M. Davies
Cellular Development	D. Garrod
Brain Biochemistry	H. S. Bachelard
Immunochemistry	M. W. Steward
The Selectivity of Drugs	A. Albert
Biomechanics	R. McN. Alexander
Molecular Virology	T. H. Pennington, D. A. Ritchie
Hormone Action	A. Malkinson
Cellular Recognition	M. F. Greaves
Cytogenetics of Man and other Animals	A. McDermott
RNA Biosynthesis	R. H. Burdon
Protein Biosynthesis	A. E. Smith
Biological Energy Conservation	C. Jones
Control of Enzyme Activity	P. Cohen
Metabolic Regulation	R. Denton, C. I. Pogson
Plant Cytogenetics	D. M. Moore
Population Genetics	L. M. Cook
Insect Biochemistry	H. H. Rees
A Biochemical Approach to Nutrition	R. A. Freedland, S. Briggs
Enzyme Kinetics	P. C. Engel
Polysaccharide Shapes	D. A. Rees

In preparation

The Cell Cycle	S. Shall
Microbial Metabolism	H. Dalton, R. R. Eady
Bacterial Taxonomy	D. Jones, M. Goodfellow
Molecular Evolution	W. Fitch
Metal Ions in Biology	P. M. Harrison, R. Hoare
Cellular Immunology	D. Katz
Muscle	R. M. Simmons
Xenobiotics	D. V. Parke
Human Genetics	J. H. Edwards
Biochemical Systematics	J. B. Harbourne
Biochemical Pharmacology	B. A. Callingham
Biological Oscillations	A. Robertson
Photobiology	K. Poff
Functional Aspects of Neurochemistry	G. Ansell, S. Spanner
Cellular Degradative Processes	R. J. Dean
Transport Phenomena in Plants	D. A. Baker
Membrane Assembly	J. Haslam

OUTLINE STUDIES IN BIOLOGY

Editor's Foreword

The student of biological science in his final years as an undergraduate and his first years as a graduate is expected to gain some familiarity with current research at the frontiers of his discipline. New research work is published in a perplexing diversity of publications and is inevitably concerned with the minutiae of the subject. The sheer number of research journals and papers also causes confusion and difficulties of assimilation. Review articles usually presuppose a background knowledge of the field and are inevitably rather restricted in scope. There is thus a need for short but authoritative introductions to those areas of modern biological research which are either not dealt with in standard introductory textbooks or are not dealt with in sufficient detail to enable the student to go on from them to read scholarly reviews with profit. This series of books is designed to satisfy this need. The authors have been asked to produce a brief outline of their subject assuming that their readers will have read and remembered much of a standard introductory textbook of biology. This outline then sets out to provide by building on this basis, the conceptual framework within which modern research work is progressing and aims to give the reader an indication of the problems, both conceptual and practical, which must be overcome if progress is to be maintained. We hope that students will go on to read the more detailed reviews and articles to which reference is made with a greater insight and understanding of how they fit into the overall scheme of modern research effort and may thus be helped to choose where to make their own contribution to this effort. These books are guidebooks, not textbooks. Modern research pays scant regard for the academic divisions into which biological teaching and introductory textbooks must, to a certain extent, be divided. We have thus concentrated in this series on providing guides to those areas which fall between, or which involve, several different academic disciplines. It is here that the gap between the textbook and the research paper is widest and where the need for guidance is greatest. In so doing we hope to have extended or supplemented but not supplanted main texts, and to have given students assistance in seeing how modern biological research is progressing, while at the same time providing a foundation for self help in the achievement of successful examination results.

J. M. Ashworth, Professor of Biology, University of Essex.

Polysaccharide Shapes

D. A. Rees

Principal Scientist,
Unilever Research Laboratory, Bedford

LONDON
CHAPMAN AND HALL

A HALSTED PRESS BOOK
John Wiley & Sons, New York

First published in 1977
by Chapman and Hall Ltd
11 New Fetter Lane, London EC4P 4EE

Typeset by Preface Ltd, Salisbury, Wilts
and printed in Great Britain at the
University Printing House, Cambridge

ISBN 0 412 13030 0

Distributed in the U.S.A.
by Halsted Press, a Division of
John Wiley & Sons, Inc., New York

Library of Congress Cataloging in Publication Data

Rees, David Allan.
Polysaccharide shapes.

(Outline studies in biology)
Bibliography: p.
Includes index.
1. Polysaccharides. I. Title.
QD321.R318 547′.782 77–422
ISBN 0-470-99098-8

Contents

1 Basic ideas about molecular shape

1.1 Shapes of biopolymers

Fundamental to our understanding of the properties of all biopolymers, is the shape in three dimensions – known as the *conformation.* As we shall see later, this can be fixed and stable or it may fluctuate continuously – depending on the particular polymer and on the conditions at the time. In either event, this shape determines how atoms and groups in the structure can present themselves to the outside – to other molecules, to solvents, to biological surfaces and so on. This in turn must determine physical properties such as solubility and binding behaviour, and biological interactions such as enzyme catalysis, hormone-receptor interactions, and nucleic acid replication. In short, molecular shape is the touchstone by which we hope eventually to understand the workings of living cells, tissues and even organisms in as much detail as we now understand the workings of a watch.

Carbohydrate molecules occur in every form of life, not only as the simple sugars and derivatives involved in basic energy metabolism, as the polymers which contribute to extracelluiar support in plants, animals and microorganisms, but also as integral components of molecules that we think of as central to the way life works, such as the genetic material, many enzymes, antibodies, hormones, and membrane proteins and lipids. The significance of the carbohydrate parts of all these structures is not yet understood in any detail but it is certain to involve interactions which depend upon three-dimensional shapes. This book is about the principles that determine the shapes of the carbohydrate parts of biopolymers and about the significance of these shapes insofar as we understand them at present.

Before we can discuss carbohydrate polymers themselves it will be necessary to outline some concepts about the shapes of molecules in general. We do this using simple non-carbohydrate molecules as examples, but slanting the discussion in a way that will be most relevant to carbohydrates in the sections following.

1.2 Conformational principles

The shapes of some parts of covalent molecules are fixed by the bonds which connect the atoms in them, apart from slight fluctuations caused by vibrations. In the hydrogen molecule, for example, the distance between the two nuclei is a fixed distance – the bond length. Another example, but this time involving three atoms, is the water molecule; here the atoms are held in a definite arrangement because the angle between the bonds – the bond angle – is fixed as well as the bond lengths. In

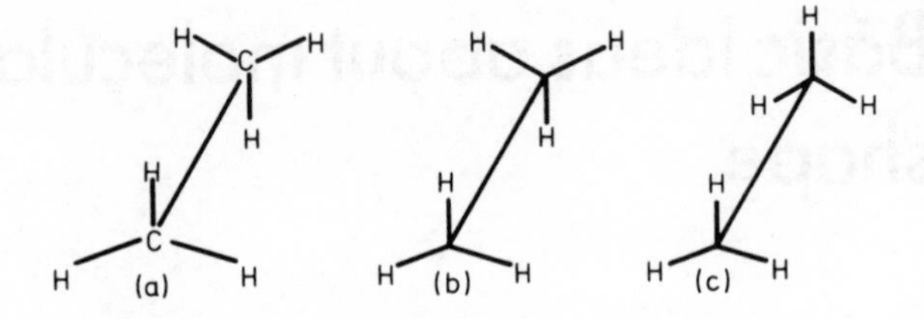

Fig. 1.1 Alternative conformations of ethane.

these molecules, the shapes are not changed by rotation about bonds.

For many larger molecules such as ethane (Fig. 1.1(a)), the shape changes with rotation about bonds. To make the geometry easier to visualize, it is usual to omit the symbols for carbon atoms ((b) and (c)). In one of these conformations (c), the H atoms are as close as they can be brought by bond rotation alone – hence it is called the *eclipsed* conformation; for similar reasons the other, (b), is called the *staggered* conformation.

The condition for stable equilibrium in a mechanical system such as a pendulum is that the total energy is a minimum. This same criterion can in principle be used to decide the most stable position when there is rotation about a bond in a molecule. Because rotation about the C–C bond of ethane alters the distances between H atoms, it must change the energy of attraction or repulsion within the molecule. The precise way in which this internal energy changes with the bond rotation is very difficult to calculate with accuracy because of the mathematical complexity of the exact treatment of interaction energies between atoms which requires the use of the methods of quantum mechanics. It is only within the last few years that the ethane problem has been satisfactorily resolved [1] and then only by the use of large computers. These calculations show, as indeed we had already known from experiment for over 30 years, that the form (b) is actually more stable than (c) by about 3 kcal mol^{-1} (12–13 kj mol^{-1}). In other words, both theory and experiment show that the energy is minimized when the C–H bonds are as widely separated as possible. The increase in energy which occurs when the bonds are moved together can be regarded as being analogous to the introduction of some form of strain into the molecule, called *torsion strain*, or *torsion energy*.

In more complicated molecules there are many other types of contribution to the total energy, and it is the balance between them that determines which conformation is most stable. One such factor is illustrated by rotation about the middle bond in *n*-butane; in this molecule there are three distinct ways in which the C–C bonds can be staggered (Fig. 1.2). Form (b) is known as the anti conformation; (a) and (c) can be regarded as mirror image forms of the same state known as the gauche conformation. The bond rotation that brings the two CH_3 groups together actually squeezes them closer than their normal distance of closest approach. The extent to which it is possible to

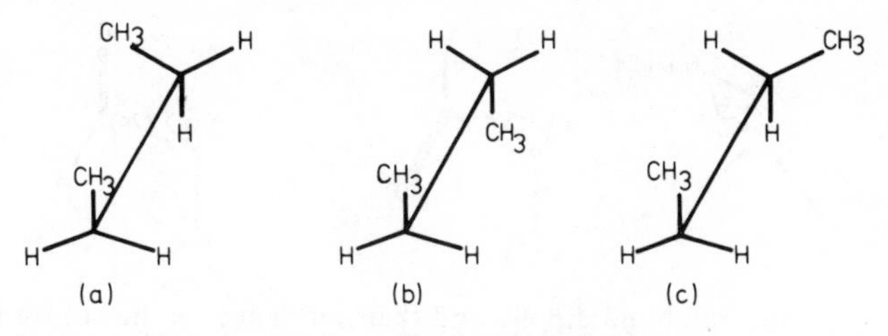

Fig. 1.2 Alternative staggered conformations of *n*-butane.

bring together atoms and groups of atoms without squeezing, can be predicted from the way that molecules are observed to pack together in crystals; we find that the 'apparent size' of an atom or group does not change very much from one molecule to another. The 'size' is usually expressed as the *van der Waals radius.* The squeezing together of CH_3 groups in butane within the usual distance of closest approach adds extra instability by *van der Waals repulsion* to the conformation in which they are eclipsed i.e. it raises the energy even further than we would expect from the torsion strain. This squeezing is even significant in the gauche form (Fig. 1.2(a) and (c)) so that it is less stable than the anti (b) by 0·9 kcal mol^{-1} (3·5 kj mol^{-1}). Incidentally, the energy differences between ethane conformations cannot be explained by van der Waals repulsion because H atoms are so small that they only just touch in the eclipsed conformation. In butane, the two CH_3 groups are said to be *non-polar* because the centres of negative and positive charge (arising from the electrons and protons which make up each group) almost coincide. For all groups of this type, attraction occurs at large distances with a force that increases as the distance diminishes until they 'in contact'. At shorter distances still, there is a repulsive interaction as already mentioned.

This idea of atomic and group size is of course merely a way of picturing the consequence of interactions between the nuclei and the electron clouds of the atoms in the structure. For *polar groups* in which the centres of negative and positive charge do not coincide, this simple picture needs to be elaborated. For a molecule which is electrically neutral overall, any separation of positive and negative centres leads to the setting up of a *dipole* – an entity in which two equal and opposite charges are separated in space. In carbohydrate molecules we are particularly concerned with the dipoles associated with oxygen atoms in the structure. In water and many other molecules, the unshared electron pairs of oxygen may be represented as occupying orbitals directed almost tetrahedrally (Fig. 1.3(a)) and therefore represent a concentration of negative charge on one side of the oxygen atom – this means that there is an *atomic dipole.* When C–O bonds are present there is also a *bond dipole* because the electrons of the bond are polarized towards the oxygen. An illustration of the influence of these dipoles on conformational stability, is in methyl formate in which the repulsion between the atomic dipole and bond dipole causes

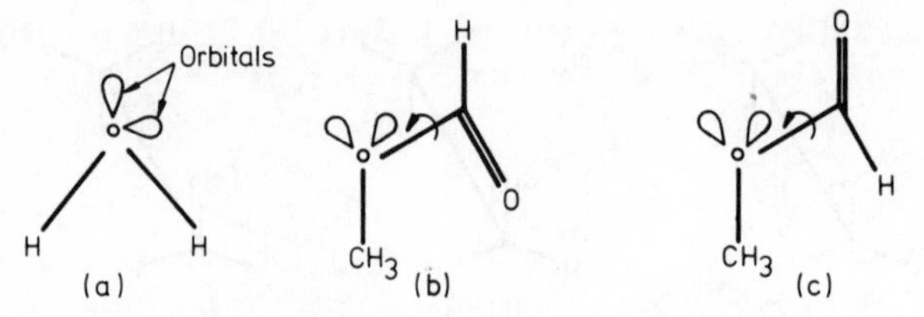

Fig. 1.3 Atomic and bond dipoles and their influence on the conformation of methyl formate.

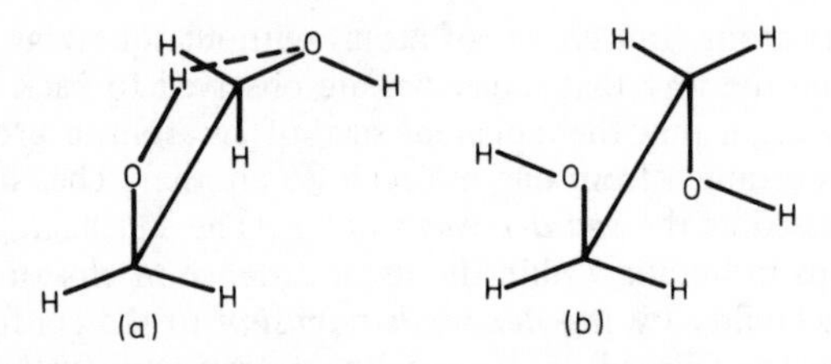

Fig. 1.4 Conformations of ethylene glycol.

the molecules to prefer the conformation shown in Fig. 1.3(b) rather than (c).

Finally, in molecules containing O–H bonds, attractions can exist between atomic and bond dipoles to set up a particularly favourable association known as the hydrogen bond:

$$O \cdots\cdots H\text{–}O$$

In ice, for example, each oxygen atom is surrounded by four tetrahedrally arranged hydrogens, of which two are bound covalently and two by hydrogen bonds. Hydrogen bonds are very important in biological molecules because they are stronger than van der Waals' attractions. Nitrogen and fluorine, as well as oxygen, can take part in strong hydrogen bonding. The effect on conformational equilibria is illustrated by ethylene glycol (Fig. 1.4). Here the oxygen atoms are close enough to allow a hydrogen bond within the molecule provided it is in the form (a) rather than (b), and there is indeed evidence that (a) gains some stability through formation of such a bond.

1.3 Shapes in equilibrium

Although we have been able to point to factors which determine the particular conformation of a molecule that will be most stable, this type of argument cannot lead to a quantitative prediction of the forms in which the molecules will actually exist. To be able to do this, we must take account of the fact that thermal collisions tend to keep conformational states mixed up. For example, although we could conclude that the anti conformation of *n*-butane is the most stable, not all the molecules in a jar of butane gas would exist in that form. This is because the 27 billion trillion or so molecules in a litre of butane gas under ordinary room conditions of temperature and pressure would be bouncing around off the walls of the container and off each other with an average speed of 6000–7000 kilometres per hour. Even if we could start with all the molecules in a single conformation they would very soon exist in a variety of conformations and states of rotation and oscillation as a result of these collisions. The possibility of the original state returning would be very remote indeed.

Consider the anti and gauche forms of butane as two distinct but interconvertible entities. On simple energy considerations alone we would expect all the molecules to exist in the form with lower energy, i.e. anti. On the other hand, neglecting energy and considering only the tendency of collisions to keep the conformations well mixed up, we would expect a statistical distribution. This would be 2:1 in favour of gauche, because there are two equivalent gauche forms (see Fig. 1.2) and the molecule therefore has twice as much chance to settle down in a gauche form. Experiments show that (at 14°C) the distribution is actually about 3:2 in favour of anti, which is obviously a compromise between the two possibilities. Although the proportion of molecules with each conformation remains the same under constant conditions, individual molecules will change continually. This means that, at equilibrium, the number of molecules disturbed from the anti conformation in a given time is exactly equal to the number that return to it from gauche, i.e. we have *dynamic equilibrium.*

One way of analysing, and indeed predicting, the positions of this and other dynamic equilibria, is through the science of *thermodynamics.* The arguments of thermodynamics show that for any closed system – which for our present discussions we may take a jar of butane gas – there is a function known as the *free energy* which plays the same part in chemical equilibrium as does potential energy in static equilibrium – that is, a chemical system will settle down in the position of minimum free energy. Under constant pressure and temperature (the usual conditions of a chemical experiment) we use the Gibbs free energy, represented by the symbol G (or sometimes, especially in North America, F). The quantity we need in order to calculate the position of equilibrium between anti and gauche forms is the *standard free energy difference*, ΔG^0 which is the increase in free energy that would occur during an imaginary change of *all* the butane molecules in the jar from the anti to gauche conformations, keeping all external conditions unaltered. The equation is:

$$\Delta G^0 = RT \log_e K \tag{1.1}$$

where R is a universal constant (the gas constant) whose value is known; T is the temperature in absolute units; K is the *equilibrium constant*, namely:

$$K = \frac{\text{Number of gauche molecules at equilibrium}}{\text{Number of anti molecules at equilibrium}}. \tag{1.2}$$

Knowing the value of ΔG^0, it would therefore be possible to predict the composition of the conformational mixture. How is this value found? It is given by:

$$\Delta G^0 = \Delta H^0 - T\Delta S^0,$$

where the functions ΔH^0 and ΔS^0 are the *standard enthalpy difference* and the *standard entropy difference* respectively, and which are defined in a similar way to the standard free energy difference, ΔG^0. Here ΔH^0 is the energy in the form of heat that would be needed to boost all the molecules in the jar from the anti to gauche conformation at a given pressure and temperature (if this could be done). This heat is stored as potential energy of repulsion in each molecule – corresponding to the 'energy bias' which tends to drive all the molecules into the anti form. The fact that the equilibrium constant does not depend on ΔH^0 alone, corresponds to our earlier conclusion that the conformation is not fixed in the position of minimum potential energy. It can be shown by theoretical arguments that the standard entropy difference, ΔS^0, which is also involved in the equation, is related to the probability factor which keeps conformations mixed up. This entropy term therefore represents the drive towards maximum probability which also influences the position of equilibrium. In principle, entropies can be calculated from measurement of heat capacities of pure substances at a series of different temperatures.

Although it may seem difficult and roundabout to use this method to derive the positions of conformation equilibria, we shall show that the basic ideas can provide a very useful framework on which to build other approaches that are simple and direct in practice. It will not actually be necessary to consider the isolation of individual conformational states for measurement of entropies and heat changes.

2 The building units

Carbohydrate building units usually have cyclic structures – which means that their component atoms are joined in a way that forms a closed ring. One of the ring atoms is almost always oxygen and the remaining atoms are carbon. To understand the shapes of the very many different units which exist in natural polymers, it is helpful to classify them according to the number of atoms involved in this ring; this number may be seven, six, five or zero (Fig. 2.1).

In polysaccharides, glycoproteins and glycolipids, the sugar units usually have six membered ring structures – and an enormous variety of these occur naturally. In DNA and RNA and a few plant and microbial polysaccharides, sugar units exist with five membered rings and, as we shall see, these shapes have very different properties. The open chain and seven membered forms are much less important in biology. In this chapter we shall discuss first the characteristic properties of each of these types of ring, then the principles which operate when, as often happens, different shapes can interconvert.

2.1 Pyranose (six membered) forms [2]

The generic name that is given to this group, derives from a class of chemicals known as pyrans, which also contain five carbon atoms and one oxygen in a six membered ring. An example, and indeed the most common example of a building unit of natural carbohydrate polymers is a six membered ring form of glucose known as β-D-glucopyranose. The way in which carbon, hydrogen, and oxygen atoms are joined together in this molecule is shown in Fig. 2.2(a). A more usual representation is shown in (b); many of the carbon and hydrogen atoms are omitted for convenience and clarity, and an attempt is made to show the geometrical shape. To refer to particular atoms, there is a convention by which the carbon atoms are numbered, shown in (c).

β-D-glucopyranose can exist in alternative conformations which convert to each other by rotation about the bonds within the closed ring. The conformation with lowest internal energy can be predicted by the following arguments based on principles described in Chapter 1. We know from the fundamental chemistry of carbon and oxygen that all bond angles in the structure will be close to tetrahedral. These bond angles can be altered only at the expense of introducing *angle strain*, which would increase the energy of the molecule and therefore the preferred conformations are those which are free from angle strain. In these, the ring cannot be flat, it must be puckered with the approximate shape of a chair or boat. Fig. 2.2(b) shows one of the two possible *chair con-*

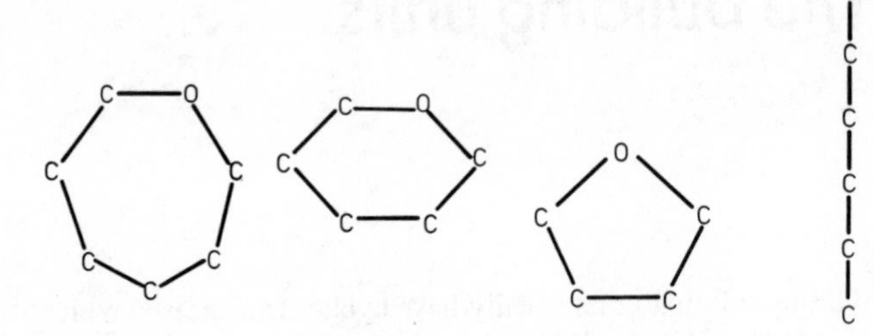

Fig. 2.1 Ring skeletons of carbohydrate building units.

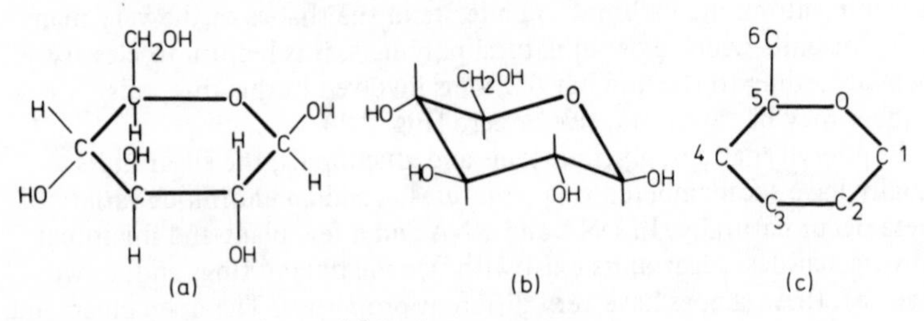

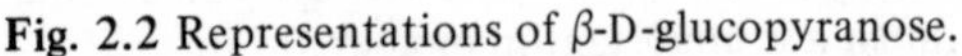

Fig. 2.2 Representations of β-D-glucopyranose.

formations. In old nomenclature this is designated the C1 conformation; in the newer and more rational nomenclature which, however, is not yet universally used, it is designated the $^4C_1$ conformation (the letter *C* signifies a chair form and the two numerals represent the ring atoms which are above and below the best plane when the ring is viewed so that the numbering appears clockwise from above). Some other conformations, including two of the possible *boat conformations*, are shown with their names in Fig. 2.3. To try to understand how one conformation converts to another, three-dimensional scale models are much better than drawings – the reader is urged to use models when following this and other parts of the book. It will then be easy to see how, for example, the 1C conformation ($^1C_4$) converts to B1 ($^{1,4}B$) by rotation about the bonds which connect C(2) – C(3), C(3) – C(4), C(4) – C(5) and C(5) – 0; C(1), C(2), and the ring oxygen of the model are firmly held and C(4) is pushed upwards to cause all the rotations simultaneously.

Inspection of drawings (Figs. 2.2 and 2.3) or, better, of molecular models, shows that all bonds are staggered in the chair conformations, whereas in both the boat forms shown the bonds attached to C(2) and C(3) are eclipsed. The internal energies of the boat forms are also greatly increased by the repulsion between groups attached to C(1) and C(4) which arises because the geometry of each boat squeezes these groups together within the normal distance of closest approach. Groups attached by bonds which project more or less vertically from chair-shaped rings may be similarly, though less seriously, cramped; those which project more nearly sideways than vertically have much more space. For this reason, it is convenient to have names to distinguish the two types of

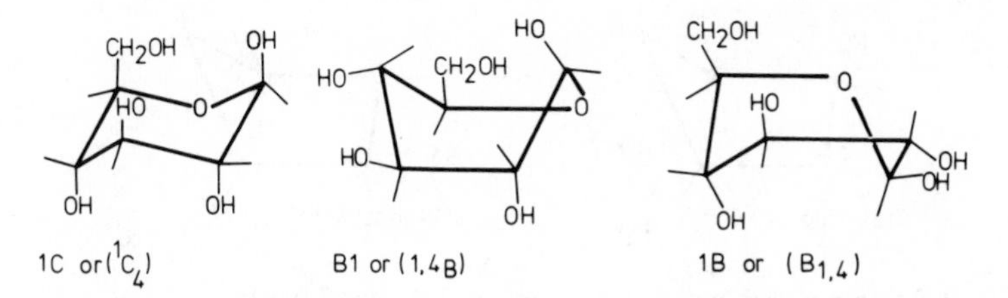

Fig. 2.3 Some of the alternative conformations of β-D-glucopyranose.

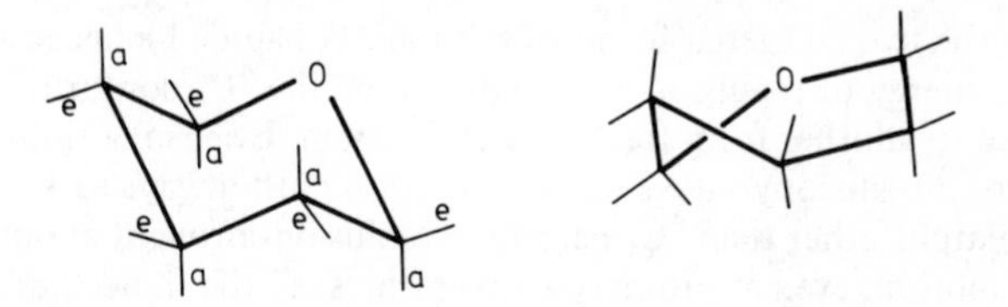

Fig. 2.4 Axial (a) and equatorial (e) bonds on the chair conformation (left); and a twist-boat conformation ($^1S_5$).

bond. The 'sideways and out' bonds in the chair conformations are called equatorial whereas the 'up bonds' and the 'down bonds' are both called axial (Fig. 2.4). It will be seen that each carbon atom carries one equatorial bond and one axial bond. Hydrogen atoms are so small that they do not repel in axial positions, although they do repel when specially close as they are at C(1) and C(4) of the boat forms shown in Fig. 2.3. Larger atoms and groups repel in all axial positions on the same side of the ring.

It should now be clear that the $^4C_1$ conformation (Fig. 2.2(b)) has a lower energy than any alternative shown because all bonds are staggered and all axial groups are hydrogen atoms which do not repel. The $^1C_4$ conformation also has staggered bonds but there are five large axial groups (four OH groups and one CH_2OH group). The boat conformations are made unfavourable by the eclipsed bonds as well as the specially unfavourable repulsions between groups attached to C(1) and C(4). Among the conformations which we have not discussed, apart from other boat forms, the only ones which are free from angle strain are *twist-boat conformations.* These are similar in shape to the boats, but twisted to allow some bond staggering and relief of repulsion – one example is shown in Fig. 2.4. However, these all have a higher internal energy than the $^4C_1$ conformation.

From the conclusion that the $^4C_1$ conformation has a lower internal energy than any of the alternatives, we would expect that β-D-glucopyranose will always tend to exist in this form and so indeed it turns out. X-ray diffraction has shown this to be the conformation in crystals, both of the simple sugar and of larger molecules in which β-D-glucopyranose is a building unit. Spectroscopic and other methods show a similar situation in solution. The energy difference between the $^4C_1$ conformation and the alternatives for β-D-glucopyranose is much larger than, for example,

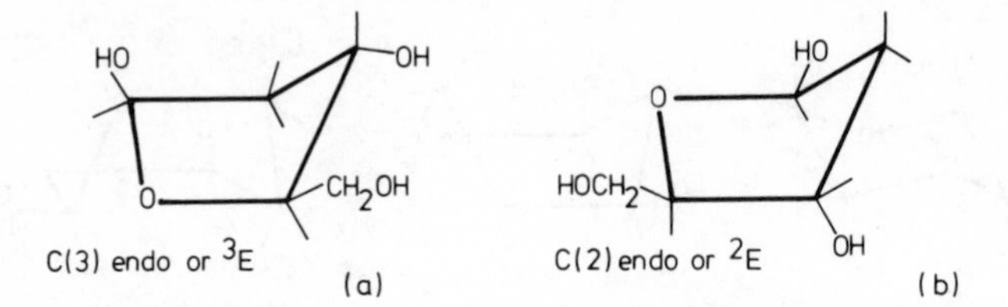

Fig. 2.5 Some possible conformations of 2-deoxy-β-D-ribose.

between the two staggered forms of *n*-butane (Chapter 1), because more than one energy of repulsion is increased when the $^4C_1$ conformation is converted to another form free from angle strain. Even so, a finite proportion of β-D-glucopyranose units in solution must always exist in conformations other than $^4C_1$ because of collisions brought about by thermal motion, even at ordinary temperatures; all the molecules are continuously changing their shapes and at any instant a proportion must occupy the higher energy states. In a later section we will show how the energy difference between conformational states can be estimated quantitatively, and how this energy difference determines the molecular distribution.

2.2 Furanose (five membered) forms [2]

Although six-membered ring forms are most commonly found in nature and as we shall see are usually the most chemically stable, the five-membered ring is also biologically important. This is often known because of its formal relationship to a group of chemicals known as the furans, as the *furanose* type. By far the most important situation in which it is found is as the building units of nucleic acids, β-D-ribose in RNA and 2-deoxy-β-D-ribose in DNA.

The important puckered forms of the five-membered sugar ring are those which, respectively, have four of the ring atoms coplanar and the fifth atom out of plane, and three of the ring atoms coplanar with two out of plane. These are known as the *envelope* and *twist* forms. It can be shown that ten envelope forms and ten twist forms are possible for each five-membered sugar ring. For 2-deoxy-β-D-ribose units in DNA, two of the envelope forms seem to be favoured, shown in Fig. 2.5(a) and (b). They are known as the C(3)-*endo* and C(2)-*endo* respectively for the obvious reason that C(3) and C(2) respectively are displaced from the plane. In the newer systematic nomenclature following the same principles as for six-membered rings, they are known as the ^3E and ^2E conformations. For comparison a twist conformation, $^3T_2$, is also shown (Fig. 2.6(a)).

Ever since the early 1950's when the conformations of DNA were being systematically investigated in fibrous specimens by X-ray diffraction, it has been known to be possible to obtain different forms which are interconvertible by changes in the salt concentration and humidity. The two main forms of DNA are known as the A and B forms. Refinement of the X-ray structures has recently shown [3] that the crucial

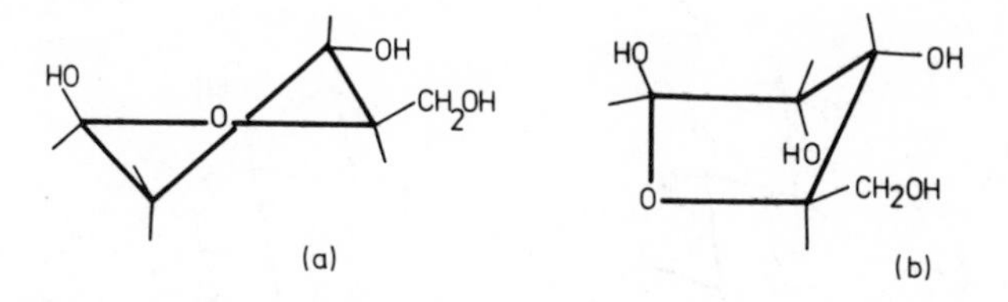

Fig. 2.6 Some possible conformations of 2-deoxy-β-D-ribose and β-D-ribose.

difference between them lies in the conformation of the deoxyribose ring, which is ^3E in the A form and ^2E in the B form. The drawings – or better the molecular models of these ring conformations (Fig. 2.5(a) and (b)), suggest that they involve rather little torsion strain and van der Waals repulsion, and are likely to be more stable than alternatives such as Fig. 2.6(a). From this inspection we would also expect that the internal energies of the ^3E and ^2E forms would be comparable: hence it is understandable that small changes in the external environment of the nucleic acid chain can cause the ring to flip from one form to another. The native state of DNA is thought to be the B form.

When we now consider the sugar unit in RNA, we see some very important differences. β-D-ribose has an extra OH group which is involved in two unfavourable interactions with gauche oxygens ($O(1)$ and $O(3)$) in the ^2E conformation, but only one ($O(3)$) in ^3E (Fig. 2.6(b)). It follows that, compared with DNA, RNA has more preference for the A rather than B backbone conformation and does not undergo the A⇌B transition. This striking difference between the two nucleic acids might be important for their biological functions.

Chains of carbohydrate units in the five-membered ring form, have a source of flexibility which is not shared by chains of six-membered chair forms. It will be recalled that the conversion of a six-membered ring from a chair to any other chair or non-chair conformation, requires the momentary increase of angle strain as one part of the ring is flattened. As a purely geometrical consequence of the way the atoms are put together – which is extremely difficult to appreciate without the use of molecular models – this is not necessary in the alteration of one boat conformation to another or to a twist-boat. It turns out that rotation about several ring bonds can occur simultaneously *without* partial flattening, to interconvert such forms. Five-membered sugar rings have a similar ability to interconvert without distortion of bond angles, and therefore they are flexible in a way that six-membered chair forms are not. In solution they can pass more smoothly and continuously between low energy forms and these forms themselves may not individually be defined as sharply as a chair form of a six-membered ring in which the precise values of ring torsion angles are fixed by the need to minimize angle strain.

2.3 Other forms [2]

The only other size of sugar ring for which conformations have been studied and characterized, is the seven-membered or septanose form.

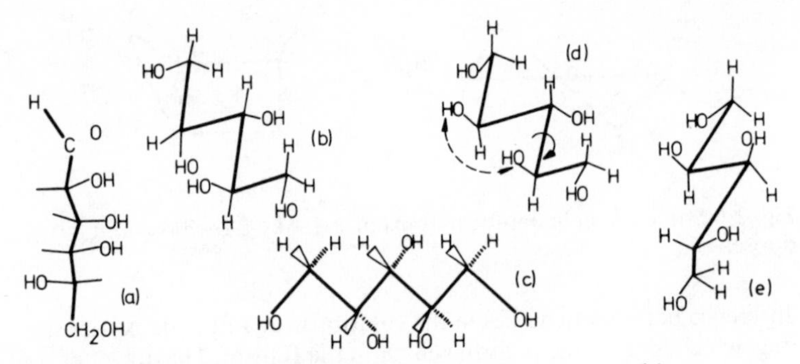

Fig. 2.7 Open chain sugars: (a) *aldehydo*-D-glucose in an arbitrary conformation; (b) and (c) D-arabinitol in alternative representations of the planar zigzag conformation; (d) and (e) ribitol in two conformations.

This exists in various chair-like, boat-like, twist-chair and twist-boat forms, all of which are flexible in the sense they are interconvertible without having to pass through intermediates with angle strain. So far, such sugar units have not been found in nature. On the contrary, special chemical procedures must be used to persuade sugars into this ring size and, given the opportunity, they will always revert to the more stable six- or five-membered rings.

More important are the *open chain* forms of sugars, such as of D-glucose (Fig. 2.7(a)). Note that this contains the same number of carbon, hydrogen and oxygen atoms as any ring form of D-glucose. All atoms are bonded in the same way to the same partners except that one bond to the ring oxygen has been removed to open the ring, and one hydrogen atom has been re-arranged to take care of the consequences for valency. This re-arrangement has created an aldehyde group $\left(-\mathrm{C}\begin{smallmatrix}\diagup \mathrm{H}\\ \diagdown \mathrm{O}\end{smallmatrix}\right)$, and for this reason the form is known as *aldehydo*-D-glucose. One way in which the sugar chain can be kept in the open form is by chemical alteration of the aldehyde group to prevent the ring closure that would otherwise occur spontaneously. In biological systems this is often done by reduction to the alcohol ($-CH_2OH$), to give a sugar derivative known as a glycitol. Examples are glucitol also trivially named sorbitol (from glucose), ribitol (from ribose), and so forth. Examples of biological occurrence are in the bacterial biopolymers known as teichoic acids in cell walls and membranes and occasionally in extracellular polymers; these may contain ribitol (or sometimes glycerol, a three-carbon glycitol) as a building unit. Ribitol is also a building unit of a limited number of capsular polysaccharides of bacteria. Glycitols occur free and in low molecular weight carbohydrates of many algae and higher plants.

The conformations of glycitol and other open chain derivatives are determined by exactly the same types of interaction energy as we have discussed for other systems. The most stable form is often the *planar zigzag* conformation shown for one member of the family in Fig. 2.7(b) and (c), which is favourable because the conformation about each C–C bond

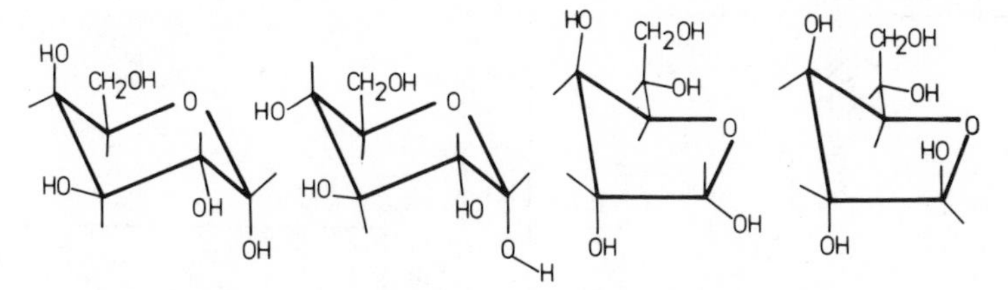

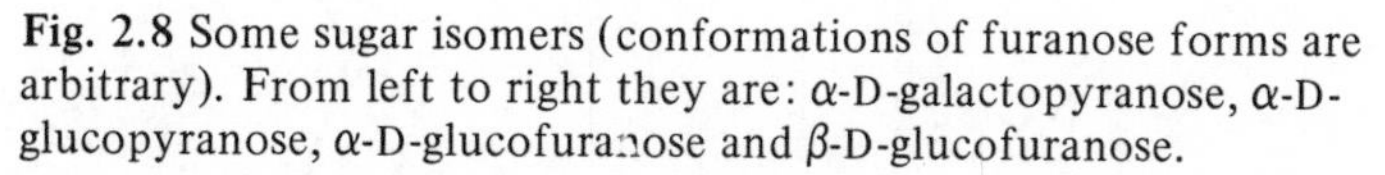

Fig. 2.8 Some sugar isomers (conformations of furanose forms are arbitrary). From left to right they are: α-D-galactopyranose, α-D-glucopyranose, α-D-glucofuranose and β-D-glucofuranose.

in the chain is staggered to place the largest neighbouring substituents as far away as possible. As the name suggests, all carbon atoms lie in one plane and the molecule has an extended shape. However, this is not always the best conformation because for some other glycitols it leads to unfavourable van der Waals repulsions between next-neighbour substituents – as for ribitol (Fig. 2.7(d)). This molecule is more stable if repulsion is relieved by rotation about the bond between C(3) and C(4), as indicated in Fig. 2.7(d), to pass into another all-staggered conformation in which the backbone is now twisted, known as a *sickle* conformation. Note that *aldehydo*-D-glucose is also unlikely to prefer the planar zig-zag conformation (Fig. 2.7(a)) because of next-neighbour interaction between the HO substituents on C(2) and C(4). Presumably a sickle conformation would again be preferred.

Open-chain sugar derivatives are expected to be very flexible in shape. There is no angle strain to resist small fluctuations of individual conformations or indeed switches of one conformation to another, and oscillations and rotations about individual bonds can occur independently. Even in the flexible rings that we have discussed above, the bond oscillations and rotations must for geometrical reasons, occur in a concerted fashion which represents some degree of restriction.

2.4 Conformation and configuration: isomers and derivatives [4]

β-D-glucopyranose (Fig. 2.2) and *aldehydo*-D-glucose (Fig. 2.7(a)) represent only two of the ways in which six carbon atoms, six oxygen atoms and twelve hydrogen atoms may be put together to construct a sugar molecule. Some other ways are shown in Fig. 2.8. Because they contain the same number of atoms of each kind, such molecules are *isomers*. It will be helpful to distinguish between sugar isomers of different types:

(i) *In configuration* at C(1). This is the difference between β-D-glucopyranose (Fig. 2.2) and α-D-glucopyranose (Fig. 2.8); and between β-D-glucofuranose and α-D-glucofuranose (both shown in Fig. 2.8).

(ii) In *configuration* at any other carbon atoms in the ring; α-D-glucopyranose and α-D-galactopyranose (both shown in Fig. 2.8) differ in this way at C(4).

(iii) In *ring size*; for example, α- and β-D-glucopyranose have six atoms joined to form a ring, whereas the glucofuranoses have five and *aldehydo*-D-glucose has no ring at all.

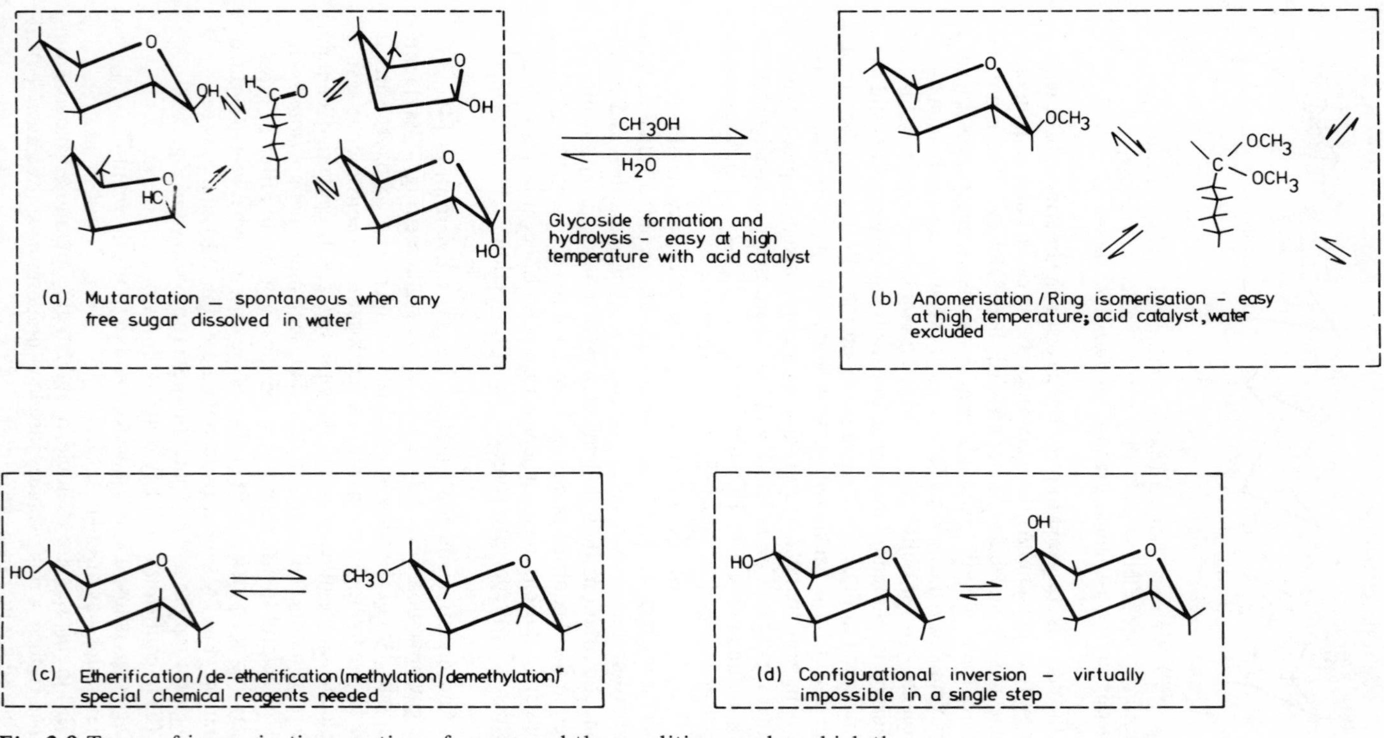

Fig. 2.9 Types of isomerization reaction of sugars and the conditions under which they occur.

The transformation of one isomer into another requires the *breaking and reformation of bonds*, not simply (as in changes of conformation) *rotation about bonds.* Since the breakage of chemical bonds usually requires more energy, interconversion of isomers is more difficult than the interconversion of conformations. It happens that the energy available from thermal collision at normal temperatures is sufficient to cause the spontaneous interconversion of only one of the types of isomer – namely that for which the essential difference lies at C(1).

In its properties, C(1) differs from other carbon atoms in the sugar ring. Changes of configuration and ring-opening isomerizations at C(1) occur in aqueous solution at room temperature, especially if a small amount of acid or alkali is added. In contrast, similar changes at other positions require specially designed sequences of chemical reactions or the action of enzymes on particular sugar derivatives. These differences may be traced to C(1) being involved in *two* carbon-oxygen bonds, whereas other carbon atoms are each involved in only one. In other words, the chemical differences are characteristic of differences between derivatives of *aldehydes* (or the related *ketones*) compared with *alcohols*. Another special property of C(1) is the way in which OH can be replaced reversibly by groups of the type OR, where R can have a variety of alternative structures. The products are known as *glycosides.*

To consider in detail why the presence of two bonds to oxygen gives C(1) these special properties, would involve a discussion of the theory of mechanisms of chemical reactions that would be inappropriate in this book – these details can in any case be found in more specialized texts [4]. The actual outcome of these influences is summarized in Fig. 2.9. This shows the interconversion between the different forms of a free sugar (by which we mean a sugar in which the substituent on C(1) is OH) is so easy that it occurs spontaneously at room temperature in aqueous solution or under biological conditions. In fact it is difficult to keep such a sugar with a single ring size and configuration at C(1) except in the crystalline state. This interconversion is known as mutarotation. For glycosides (Fig. 2.9(b)), changes in configuration at C(1) (known as *anomerization*) and ring size are easily brought about in the laboratory by heating in the presence of acid, but do not occur at biological temperature and pH unless catalysed by an appropriate enzyme. Similar conditions of high temperature with acid catalyst are required to convert glycosides to free sugar and vice versa; the position of equilibrium is determined by the relative amounts of water and the alcohol ROH. Conversion of OH to OR at any position on a sugar other than C(1) (Fig. 2.9(c)), is still more difficult and needs special reagents such as methyl iodide with sodium hydride in dimethyl sulphoxide as solvent for the forward reaction, and anhydrous boron trichloride for the reverse direction. Finally, configurational inversion at any position other than C(1) (Fig. 2.9(d)), to be achievable in a simple chemical reaction, requires such violent conditions that total destruction of the carbohydrate would occur first.

These basic reactions of carbohydrates have been described in some

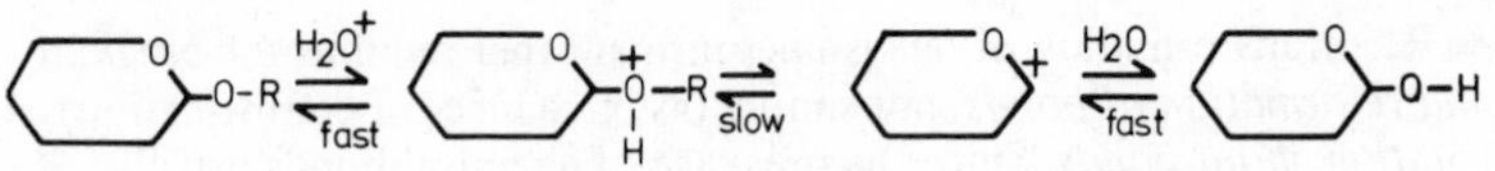

Fig. 2.10 A mechanism for glycoside hydrolysis shown with simplified formulae.

detail because they are fundamental to the behaviour of carbohydrate units in biopolymers and to the way in which these biopolymers may be handled in isolation and investigation.

2.5 Sugar shapes in hydrolysis of carbohydrate chains [5]

The breakdown of carbohydrate chains, whether in laboratory reactions of the type described in the preceding section or in normal biological processes, usually proceeds by a pathway shown in Fig. 2.10. The evidence for this comes from detailed comparisons of the kinetics of reaction for different sugar derivatives.

In the laboratory, hydrolysis is with hot dilute acid; the H^+ ions equilibrate between the oxygen atoms in the system, including those of water and glycoside, so that there is an equilibrium concentration of protonated glycoside. This decomposes slowly to a *carbonium ion*, which is so named because it has a positive charge on carbon. This ion reacts rapidly with water to give the sugar, and the overall result is therefore that the substituent 'R' – which may be the remainder of a carbohydrate chain – is split away. The carbonium ion has a high energy because the positive charge on carbon is an unstable arrangement – only six electrons are present in the outer shell instead of the usual eight. To some extent, the positive charge can be stabilised if it is 'shared' with the ring oxygen, but this needs a geometrical adjustment so that the appropriate orbitals can become aligned. This requires that C(2), C(1), the ring oxygen, and C(5) become coplanar – there is a change from the chair conformation to the so-called 'half-chair'. Inspection of models shows that this diminishes any interactions between axial groups but causes partial eclipsing of groups at each end of the C(2)–C(3) and C(4)–C(5) bonds. This knowledge enables us to predict the effect of any change in the structure of a glycoside on the energy of formation of the carbonium ion. Since this ion represents the state of highest energy in the entire sequence (Fig. 2.10), lowering its energy speeds up the reaction, and raising the energy slows the reaction. Thus we can understand why sugars which carry an electron attracting group such as –COOH or $-NH_3^+$ have glycosides which are hydrolysed relatively slowly; these groups make the positive charge even less stable. Likewise, glycosides of xylopyranose are hydrolysed faster than those of glucopyranose because the substituent on C(5) is H instead of CH_2OH (see Chart 1). Not only does the H have less tendency to withdraw electrons, but it is also smaller and therefore does not impede the conversion from chair to half-chair.

In biological processes, the breakdown of carbohydrate chains is brought about by enzyme catalysis. This speeds the process enormously and permits it to take place under normal physiological conditions. If the reaction follows the usual pathway (Fig. 2.10), the enzyme must in some way bring about a dramatic reduction in the free energy difference between the reactants and the carbonium ion; or, more precisely, between the enzyme-bound forms of these. One glycoside-splitting enzyme has been submitted to sufficiently detailed structural and mechanistic studies to show that this is indeed so. This is lysozyme which splits the bacterial cell wall peptidoglycan (Section 4.2.4). The active site in which the substrate is bound to undergo reaction, carries at least three features which assist formation of the carbonium ion. These are, (i) the COO^- group of an aspartate side chain is held close to the carbon atom on which the positive charge must develop, so providing some electrostatic stabilization,(ii) the COOH group of a glutamic acid side chain is poised close to the glycosidic oxygen to facilitate the transfer of a proton, (iii) in the process of binding the substrate to the enzyme surface, the sugar unit is distorted from its usual chair form towards the half-chair, so that even before it has formed the carbonium ion it has moved towards the necessary shape.

2.6 Prediction of shapes

Having seen in preceding sections that sugars can exist as mixtures of interconverting isomeric and conformational forms, it is natural to ask now whether the compositions of such mixtures are at all predictable. Since the common forms are α- and β-pyranoses in their chair conformations, the problem will be discussed here for mixtures of these.

Two methods of prediction have been successfully used which are quite different but complementary in approach. The first is entirely empirical and starts from measured equilibrium constants to derive the corresponding free energies (see Section 1.3) and then to make use of trends and regularities in these values. The second method attempts to build up a complete picture of all the actual attractions and repulsions between the atoms present. This approach is clearly more general and is also more satisfying in that it relates the behaviour of the sugars to fundamental physical forces. However, its disadvantage is that its application is much more complex and time consuming than the first method. There are assumptions in both methods which need to be checked.

2.6.1 Estimation of interaction free energies [5]

This approach is guided by the arguments outlined above, which lead us to expect that energy differences between sugar forms can be attributed mainly to the particular interactions shown in Fig. 2.11. In reducing the problem to these six terms, it is assumed that (i) the interaction energy between adjacent substituents as in C_1:O_2 and O_1:O_2 (Fig. 2.11) would be the same for an axial-equatorial relationship as for the equatorial–equatorial situation shown (this is justified by inspection of molecular models which shows that the separation distances are similar); (ii) the interaction energy depends only on the atom that is directly bonded to the ring atom, for example CH_3O– and HO– are energetically equivalent

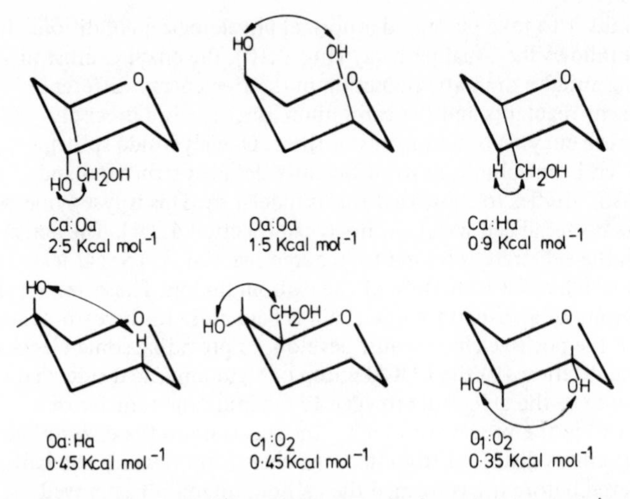

Fig. 2.11 Important repulsions between substituents on sugar rings, and estimates of the free energies associated with them.

As described in the text, extra terms are required when considering configurational equilibria at C(1): an axial oxygen on C(1) is stabilized relative to equatorial by 0.55–1.00 Kcal mol^{-1}, depending on the disposition of groups on C(2) and C(3).

and likewise for CH_3– and $HOCH_2$– (justified because atoms further away are expected to rotate from unfavourable repulsions). The six basic interaction energies of recurring importance on sugar chairs (Fig. 2.11) are thus C:O, O:O, C:H, and O:H interactions across axial:axial positions, and C:O and O:O interactions across equatorial:equatorial or axial:equatorial positions. We can of course think of sugar molecules in which none of these interactions would exist; these are considered to have the same 'baseline' free energy. If each of the six fundamental interaction types (Fig. 2.11) always changes the free energy of a sugar ring by the same amount whenever and wherever it is present, we only need to know the free energy increment associated with each type in order to calculate the total free energy of any particular sugar relative to this baseline – all we have to do is to add up the relevant terms. Hence, the proportions of alternative sugar forms in an equilibrium mixture, is easily found from the free energy *difference* between these forms via equation (1.1). For example, a free energy difference of 0.4 Kcal mol^{-1} would lead to an approximately 2:1 mixture of forms, and 1.3 Kcal mol^{-1} to approximately 10:1.

The free energy contributions associated with each of the six basic interactions, were derived by using essentially the same argument in reverse. Measurements were made of the equilibrium compositions of mixtures of a number of known systems – actually conformational equilibria for sugar-like cyclohexane derivatives known as inositols and mutarotation equilibria for pyranoses (Fig. 2.9) – for both of which it happens that convenient and accurate experimental methods are available. Then it was possible to write down and solve a large number of simultaneous equations in which the six basic free energy terms were the unknown quantities. The values that emerged are shown in Fig. 2.11.

This entire argument assumes that the basic interaction types are identified correctly and that the six-membered chair geometry does not change significantly from one sugar derivative to another to change the distance between substituents and hence the interaction free energy. In fact, it turns out that a correction term is needed for successful predictions of mutarotation equilibria because α forms have a source of stability relative to β which cannot be explained in terms of the analysis given above. From other evidence (such as the alternative analysis described below) this is believed to arise mainly from bond dipole interactions within the sequence of atoms C(5)–O–C(1)–O. This correction is known as the 'anomeric effect' and values are given in the legend to Fig. 2.11. When this is incorporated, the method gives excellent predictions of the equilibrium compositions of sugar forms.

2.6.2 Calculation of conformational energies [6]

A second and quite different approach to the prediction of sugar shapes is to start from as rigorous a treatment as possible of interatomic forces. In principle this would use the exact methods of quantum mechanics but this is not feasible at present and the approximate methods of quantum mechanics have not so far given very good results. More feasible and reliable is the evaluation of such terms as van der Waals attraction and repulsion, hydrogen bonding, polar interactions and bond torsion, from the interatomic distances in the molecule. It is also necessary sometimes to consider more fundamental variation of molecular geometry, such as the angle strain that is introduced when the molecule is distorted. All this requires the use of functions for the variation of energy with interatomic distance. These functions have been checked and calibrated against fundamental measurements such as the energy in the form of heat that is required to separate small molecules of known geometry. For example, the heat of sublimation of *n*-hexane has been used in establishing functions for the van der Waals interactions involving carbon and hydrogen.

In this method, we deal with *potential energies* which correspond closely to the ΔH terms of equation (1.2) rather than the free energies. A sugar molecule such as β-D-glucopyranose is treated as follows. The first step is to list the stereochemical constraints that can safely be assumed, such as values of bond lengths and preferred values for bond angles. Then the remaining parameters are varied systematically; for example, for the $^4C_1$ form of β-D-glucopyranose this could involve the variation of those torsion angles indicated in Fig. 2.12. More thorough treatments are possible and perhaps desirable in which the torsion angles between ring atoms and also the bond angles themselves are varied but the basic principles are essentially the same. We start with a set of arbitrary values for the conformational variables and calculate the internal energy as the

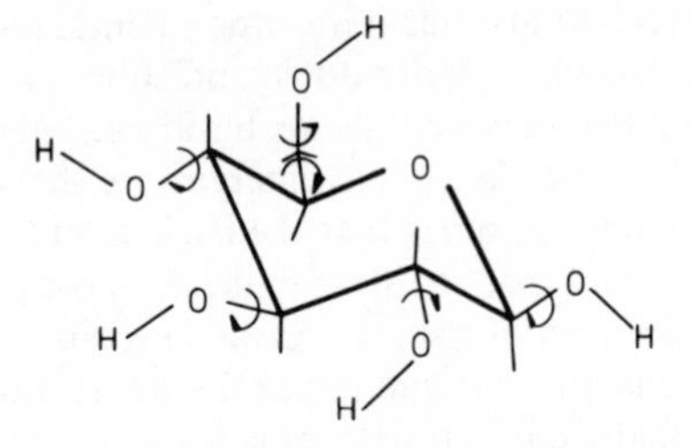

Fig. 2.12 Conformation variables in the $^4C_1$ form of β-D-glucopyranose.

sum of the van der Waals, polar, hydrogen bonding and torsion contributions, each of which is evaluated from interatomic distances by means of appropriate energy functions. The mathematics is elementary but so lengthy that it is only feasible by the use of an electonic computer. The value of one torsion angle is then changed by a small increment (say 10°), after which the relevant atomic positions and hence the interatomic distances and internal energy are recalculated. This is repeated after many further changes of conformational variables, until all possible geometrical relationships have been explored thoroughly and systematically. From this survey we can identify the set of values for the conformational variables which gives the minimum potential energy, or, as is actually more appropriate for prediction of states in solution, a special kind of average energy, known as the *thermodynamic average.* In solution, thermal collisions prevent molecules from settling down in the conformation of minimum energy; the molecules are mobile but tend to spend more time in states of lowest energy and the thermodynamic average is weighted to allow for this.

So far we have neglected the consideration of entropy, even though this can in principle have an important influence on equilibrium (Section 1.3). It is beyond the scope of this book to discuss how this is calculated and in any case it turns out that the entropy differences between chair forms of sugars are small and indeed, insignificant; so that the major influence on equilibrium positions is the enthalpy term i.e. the potential energy which is calculated by the method described. This is not necessarily true for other forms, however, and entropy is certainly relevant to polymer conformations as we shall see later.

These calculations are now refined to the point where it seems they can usually predict that conformations and configurations that will be present in a sugar equilibrium and the proportions, and may even predict minor distortions and deviations from idealized chair forms as when the ring flattens to some extent and acquires angle strain to relieve unfavourable repulsions. It emerges that the anomeric effect is included implicitly in this approach because no extra correction turns out to be necessary. One important qualification must, however, be made in discussing the reliability of these calculations. They have considered the interactions within a single molecule in isolation; the nearest counterpart in physical reality would be a sugar molecule in the gas phase at low pressure whereas, of course, to understand relationships in biology we need to understand sugar molecules surrounded by water molecules. Interactions with water must involve extra energy and entropy terms which we have not yet even begun to consider but the predictions correspond well with experimental measurement for most sugars and we conclude that these terms usually cancel when we consider free energy *differences* which is all that is needed for our purposes (equation (1.1)). Some exceptions do exist, especially for sugar forms with several adjacent equatorial hydroxy groups, such as β-D-glucopyranose which has four such groups (Fig. 2.2). It may be no coincidence that the spacing and arrangement of these adjacent oxygen atoms matches closely the situation that is believed to exist in ordered clusters of water molecules so that a favourable 'lock and key' interaction could perhaps occur to give the extra stability which the calculations fail to predict.

2.7 Natural building units

The sugar units in a biopolymer are usually joined by covalent bonds which connect C(1) through an oxygen atom to a carbon atom of the next part of the biopolymer structure which is often another carbohydrate unit but may also be an amino acid or lipid unit. This is the same kind of linkage that occurs in glycosides, and it has all the same properties (Section 2.4). Long chains of sugar units can be built up in this way in which each unit has a fixed ring size and configuration at C(1). Analogous linkages, in which a nitrogen atom rather than oxygen provides the bridge from C(1) are also known, for example in nucleic acids and many glycoproteins.

Some of the more important building units in nature are shown in Chart 1. Many are related in a fairly simple way to glucose – for example galactose and mannose only differ from glucose in configuration at a single carbon atom each, and some others are formally derived by simple modification of these three sugars. The CH_2OH group may be replaced by COOH (e.g. glucuronic acid), CH_3 (e.g. fucose), or H (e.g. xylose); and an OH group may be replaced by NH_2 (e.g. galactosamine). All these units occur commonly with pyranose (i.e. six-membered) rings, and are sometimes modified further by substitution (Chart 1). Some examples of polymer building units with furanose (i.e. five-membered) rings are also known. A particular carbohydrate polymer may contain only one kind of sugar unit or it may contain several.

Like most fairly complicated biological molecules, sugars are dissymmetric. This means that they may exist in left-handed and right-handed forms which are mirror images but not identical with each other, known as the L- and D-sugars (e.g. α-D-galactopyranose and α-L-galactopyranose). Most building units in nature are D-sugars, apart from L-rhamnose and L-fucose. L-galactose occurs occasionally, but is less common than D-galactose. Although arabinose usually occurs as the L-sugar, this form is actually related biosynthetically to the D-forms of other sugars.

Building units usually, but not always, contain either five or six

Chart 1: Some important carbohydrate units and derivatives.

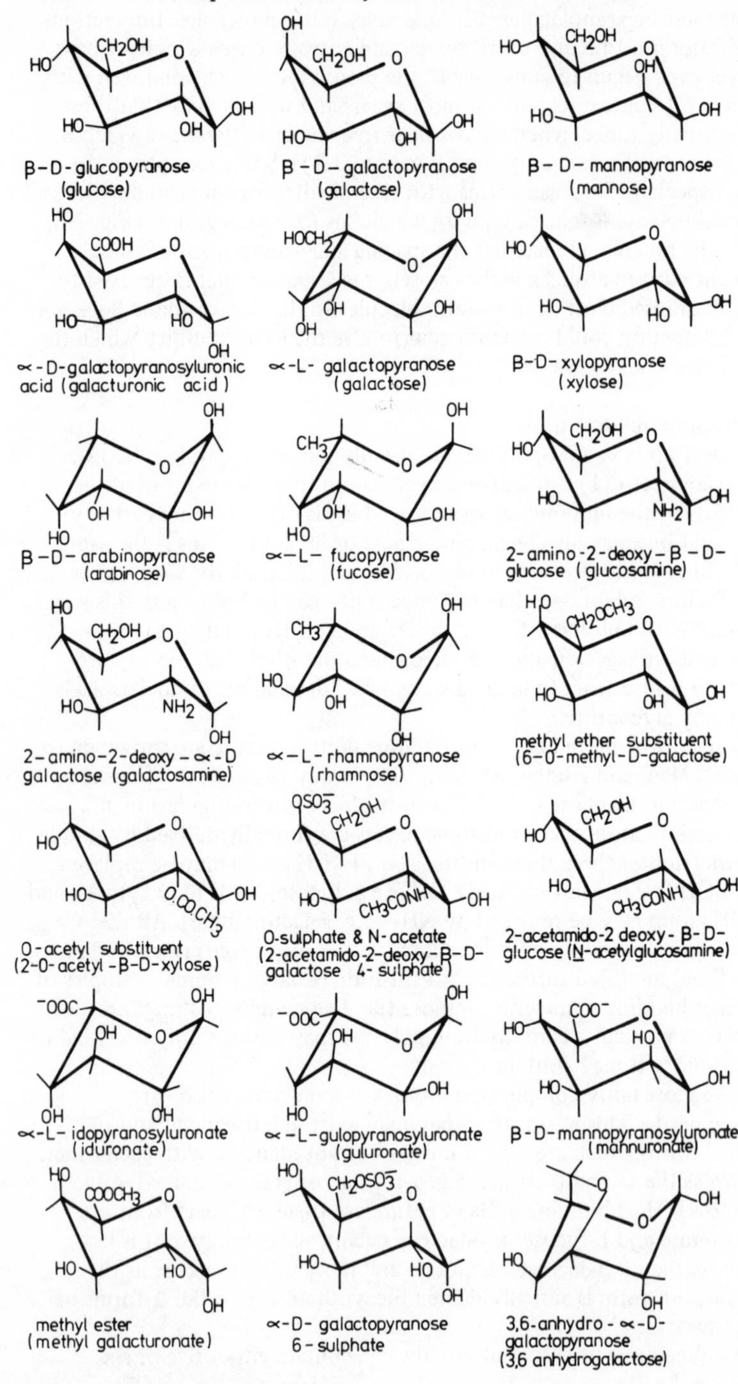

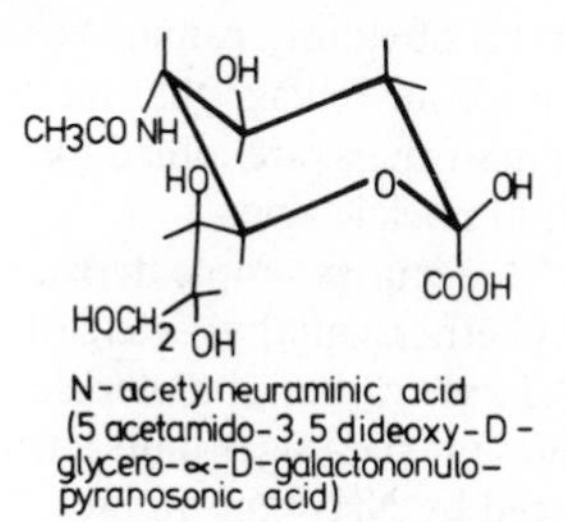

N-acetylneuraminic acid (5 acetamido-3,5 dideoxy-D-glycero-∝-D-galactononulopyranosonic acid)

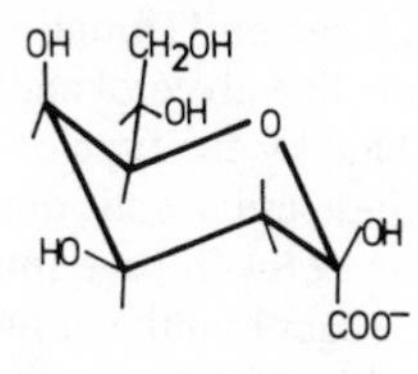

2-keto-3 deoxy-β-D-manno-octonic acid (2-keto-3 deoxy-octonic acid or "KDO")

β-D-fructofuranose (fructose)

β-D-galactofuranose (galactose)

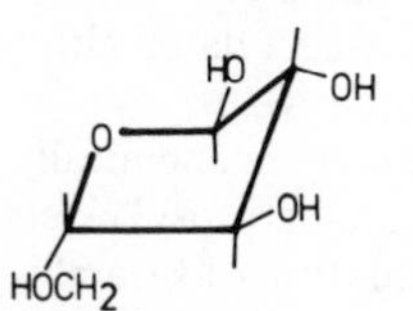

∝-L-arabinofuranose (arabinose)

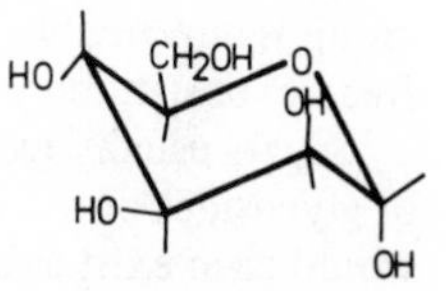

∝-D-mannopyranose

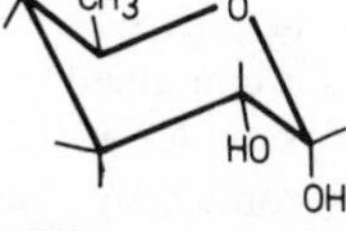

∝-abequose (3,6-dideoxy-∝-D-galactose)

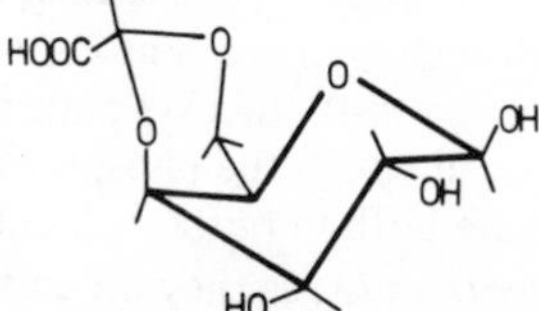

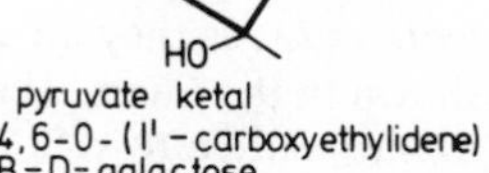

pyruvate ketal 4,6-O-(1'-carboxyethylidene) β-D-galactose

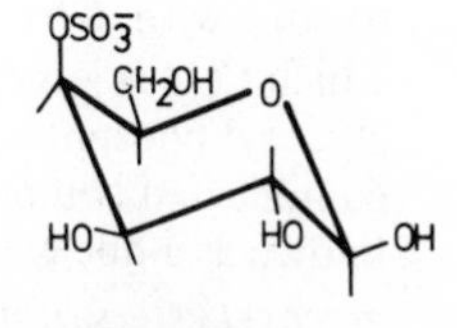

β-D-galactopyranose 4-sulphate (galactose 4-sulphate)

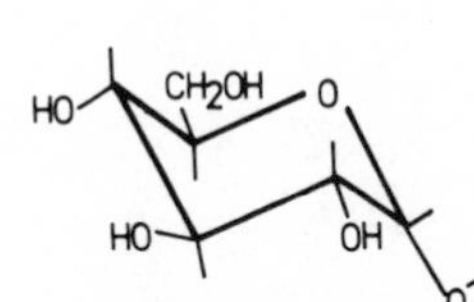

∝D-glucosyl phosphate

'Uridine diphosphate glucose' (UDPG)

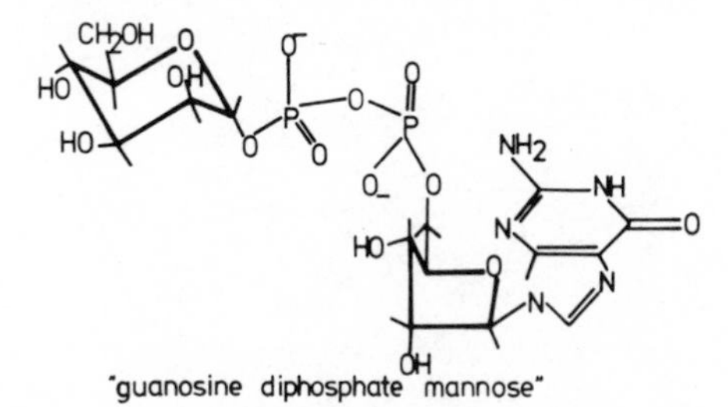

"guanosine diphosphate mannose"

carbon atoms. Neuraminic acid (very common in carbohydrate polymers in animals) contains nine; glycerol (found in certain bacterial polymers) contains only three. Units like glycerol and ribitol obviously cannot be joined in the polymer chain by the type of connecting linkage that we have described above. These units, and sometimes others, are joined by phosphate bridges similar to the linkage found in nucleic acids.

As shown in Chart 1, sugar building units also occur as simple derivatives such as sulphate and acetate esters, methyl ethers, methyl esters of carboxylic acid functions, acetals or ketals of simple carbonyl molecules such as pyruvic acid. When amino sugars are present (i.e. sugar units such as glucosamine in which an OH has been replaced by NH_2) the amino group is almost always *N*-acetylated although it does occasionally occur free or substituted in other ways.

Sugars usually exist and are chemically changed in nature in the form of glycosides or the related glycosyl esters. Unlike free sugars, which would each exist as a mixture of several isomers under biological conditions, these compounds have the ring size and configuration at C(1) both fixed and the biological machinery therefore has only one isomer of each to cope with. *Glycosyl phosphates* are a special type of glycosyl ester in which C(1) is joined through an oxygen atom to phosphorus. Simple glycosyl phosphates, such as α-D-glucosyl phosphate (Chart 1), and more complicated compounds in which the phosphorus is contained in a unit known as a nucleotide are both important. Some examples of these *glycosyl esters of nucleotides* or, as they are more commonly called, *sugar nucleotides* are shown in the Chart. The nucleotide part of each structure is itself a *pyrophosphate ester of an N-glycoside*, which is not changed in the biological interconversions and polymerizations of sugars although it does seem to be essential as a carrier in many key pathways.

3 The linkages

3.1 Linkage structures and patterns

The simplest structures in which the units are joined in the same way as they are most commonly found in biopolymers, are *disaccharides*, such as cellobiose and maltose. Conformational formulae (Fig. 3.1(a) (b)) give the most complete representation; but sometimes when discussing structure and configuration it is convenient to leave aside the question of conformation and use Haworth formulae (Fig. 3.1(c) (d)).

The connecting linkage in cellobiose is from C(1) to C(4), and the configuration at C(1) in the linkage is the same as in β-D-glucopyranose (Fig. 2.2(b)); the two units are therefore said to be joined by a *β-1,4 linkage.* In maltose the linkage is *α-1,4.* Other D-glucopyranose disaccharides exist with linkages which connect to C(1), C(2), C(3), or C(6), with either α or β configuration – hence there are ten possibilities. Most of these are found as segments of natural polymers. In addition, each of the many other natural sugar building units (e.g. Chart 1, pp. 28–29) can in principle form a similar variety of disaccharides. When we consider mixed units and/or linkages, especially with longer chains, more and more possibilities arise because of the alternative sequences. Thus, simple variations on a single way of putting sugar units together can generate a large family of different structures. Not all of these actually occur in nature, but a large number of them do and it is the main purpose of this book to examine the overall shapes and interactions and the properties that result from them. For this purpose, it will be helpful to classify the variety of covalent structures into three groups:–

- I *Periodic sequences* – in these, the sugar units are arranged in a repeating pattern along the chain. Examples are amylose, cellulose and hyaluronic acid (Chart 2). Quite complicated repeating sequences of sugar units might be present as, for example in the O-antigen chains of Gram-negative bacteria in which four (or sometimes even five) sugars may be combined in each repeating unit (Chart 2).
- II *Interrupted sequences* – these chains also have repeating sequences but separated, or interrupted, by departures from regularity. Examples are in alginates, carrageenans and certain glycosaminoglycans (Chart 2).
- III *Aperiodic sequences* – these are characterized by irregular sequences of sugar units, linkage positions and sometimes configurations. Often the structure is made up of a number of shorter chains of sugar units so that the overall pattern is branched. Two examples are shown in Chart 2.

Chart 2: Examples of the types of sequence of sugar units in carbohydrate chains. (For formulae of sugar units themselves, see Chart 1, pp. 28–29.)

Name of Polymer	*Occurrence and function*	*Sugar unit present and symbols**	*Sequence**
Periodic type			
Amylose	Almost all higher plants as energy storage material.	4-linked α-D-glucopyranose (○).	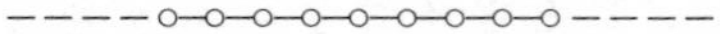 – – – – ○–○–○–○–○–○–○–○–○ – – – –
Cellulose	Almost all plants, as the framework for cell walls.	4-linked β-D-glucopyranose (●).	 – – – – ●–●–●–●–●–●–●–●–● – – – – –
Hyaluronate	Almost all animals, as extracellular gel which contributes to hydration, physical protection and perhaps control of differentiation.	4-linked β-D-glucopyranosyluronate (□); 3-linked 2-acetamido-2-deoxy-β-D-glucopyranose (■).	 – – – – □–■–□–■–□–■–□–■ – – – –
O-antigen	Carbohydrate chains attached to lipid which is embedded in the outer membrane of a species of *Salmonella.* Related structures in other Gram-negative bacteria. Antigenic properties.	α-abequose (○); 2-linked β-D-mannopyranose carrying side chain through position 3 (●); 4-linked β-L-rhamnopyranose (□); 3-linked α-D-galactopyranose (■).	 – – – ●(○)–□–■–●(○)–□–■–●(○)–□–■–●(○)–□–■ – – –
Interrupted type			
Alginate	Extracellular and intercellular gels in certain algae, contributing to hydration, ionic relationships, physical form and protection.	4-linked α-L-gulopyranosyluronate (○); 4-linked β-D-mannopyranosyluronate (●).	 –○–○–○–○–○–○–○–●–●–●–●–●–●–●–●–●–○–●–○–●
ι-carageenan		3-linked β-D-galactopyranose 4-sulphate (□); 4-linked 3,6-anhydro-α-D-galactopyranose 2-sulphate (■); 4-linked α-D-galactopyranose 2,6-disulphate (△).	 – – – □–■–□–■–□–■–□–■–□–△–□–△–□–■

Pectin†	Component of cell plate and cell wall of many growing plant tissues, providing hydrated and deformable matrix to permit shape change during growth; cementing substance between walls of mature plant tissues.	4-linked α-D-galactopyranosyluronate (○) and its methyl ester (●); 2-linked L-rhamnopyranose (▲).	●—●—●—●—●—●—●—▲—●—●—○—○—○—●—
Dermatan sulphate	Connective tissues of animals, especially tough tissues such as skin and heart valve; contributes to strength and elasticity, probably through ion and water relationships and protein binding.	4-linked β-D-glucopyranosyluronate (□); 4-linked α-L-idopyranosyluronate (■); 3-linked 2-acetamido-2-deoxy-β-D-galactopyranose 4-sulphate (○).	■—○—■—○—■—○—■—○—■—○—□—○—□—○—□—○—■—○
Aperiodic type			
Carbohydrate chain of ganglioside–G_{IV}	Member of a family of carbohydrate chains attached to lipid which is anchored in membrane of animal cells; they seem to function as receptors for certain viruses and toxins and other interactions of cell with environment.	Terminal and 9-linked *N*-acetyl-α-neuraminic acid (□); 3-linked β-D-galactopyranose (●); 3-linked 2-acetamido-2-deoxy-α-D-galactopyranose (○); 4-linked β-D-glucopyranose (▲); 4-linked β-D-galactopyranose, carrying side chain through to position 3 (△).	□—●—○—△(—□—□)—▲
Carbohydrate chain of an immunoglobulin IgG	Chains of this general type are attached to the 'stem' of most immunoglobulins; possibly they modify interactions between domains and/or the conformation change following binding of antigen.	4-linked 2-acetamido-2-deoxy-β-D-glucopyranose (●); 2-acetamido-2-deoxy β-D-glucopyranose with incompletely characterized linkage position (⊕); 3-linked α-D-mannopyranose, carrying side chain through position 6 (△); 2-linked α-D-mannopyranose (▲); 6-linked or terminal 2-acetamido-2-deoxy β-D-glucopyranose (○); terminal α-L-fucopyranose (□); terminal β-D-galactopyranose (■).	■—○—▲—△(—▲—○)—⊕—●(—□)

*Note that symbols may have different meanings for different entries.
†The backbone structure only is shown for this polymer – typically, this carries side chains of other sugars in addition to the features shown.

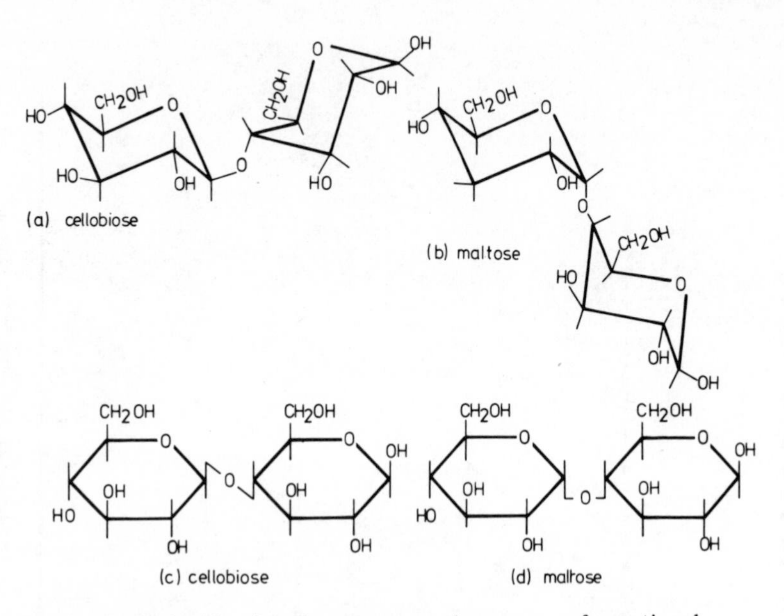

Fig. 3.1 Disaccharide of D-glucopyranose shown as conformational formulae (above) and Haworth formulae (below).

Many (but not all) carbohydrate chains are combined in larger structures which contain non-carbohydrate parts as well, especially protein or lipid. The term *polysaccharide* or *glycan* is often used to refer to chain molecules built entirely from carbohydrate units. *Glycoproteins, proteoglycans* and *peptidoglycan*, are various types of complex molecule which contain carbohydrate and peptide units. *Glycolipids* and *lipopolysaccharides* are similarly based on carbohydrate and lipid. Our classification in terms of the sequence pattern, is useful for the carbohydrate chains of all these categories.

3.2 Linkage conformation [7, 8, 9]

Consider first a simple disaccharide such as cellobiose. Although we know from the arguments of Chapter 2 that each glucose unit in this structure would exist in the $^4C_1$ form which is rigid and well defined, this does not settle the overall conformation because rotations are possible about the two bonds to the bridging oxygen. The two angles of rotation (see Fig. 3.6) are generally designated by the Greek letters ϕ and ψ. For the reasons discussed in Chapter 1 for simpler molecules, thermal collisions might perpetuate oscillations and rotations about these bonds to make the overall shape mobile rather than fixed. The overall shape would then have to be defined in terms of an equilibrium between alternative states of rotation which might correspond to different internal energies.

Of the two approaches to conformational prediction (Section 2.6), only the method of direct calculation from potential energy functions is feasible here. To work through any alternative based on the estimation of interaction free energies, we would first need to identify a small number of separate, discrete conformations which we would suppose to be the states in equilibrium – presumably these would correspond to the six staggered forms. We shall see, however, that to understand the specific interactions that are likely to be important biologically, we need to define not only the energy minimum within which the molecule is located, but *whereabouts* in that minimum. This will require knowledge of the detailed way in which the internal energy changes continuously with the angle of rotation. This is precisely the type of information that the method based on the use of potential functions aims to provide.

To understand the conformation in a disaccharide such as cellobiose, we would like to have a graph showing the change in potential energy with rotation about the C(1)–O and O–C(4) bonds. To derive such a graph, the conformation energy would first be calculated for a model in which ϕ and ψ were first given arbitrary values, by adding up the energies of attraction and repulsion between all pairs of atoms much as we described for monosaccharide shapes in Section 2.6.2. These energies would be calculated from the functions for the relation between terms such as van der Waals attraction and repulsion, hydrogen bonding, and polar interactions, and the distance of separation. One bond to the bridge oxygen (say ϕ) would then be fixed while the model was rotated in increments through a complete revolution about the other. The new atomic positions would be recalculated after each increment and used to derive each new conformational energy. The process would then be repeated about the other bond (i.e. to change ϕ) *for each increment* of rotation about the first. In this way, all possible conformations of cellobiose would be sampled. Because the conformational energy corresponding to a particular value of ϕ depends also on the value chosen for ψ and vice versa, the energy cannot be represented by a simple graph but has to be shown as a contour map – a surface with bumps and hollows like mountains and valleys rather than a graph with peaks and troughs. The results for one of the most detailed calculations yet carried out for a disaccharide, is shown in Fig. 3.2. The deepest hollow is expected to represent the preferred conformation. This can be checked against the crystal structures of cellobiose derivatives that have been characterized by X-ray diffraction, to test whether the values of ϕ and ψ to which these known forms correspond do indeed fall in this region. It turns out (Fig. 3.2) that the different derivatives do not crystallize with exactly the same values of ϕ and ψ – presumably each adjusts its shape in a different way to pack efficiently in the crystal – but always they are close together and within the predicted boundary. Fig. 3.3 shows the form that is predicted by these calculations to have the minimum energy and from which all the other forms can be considered to be derived by small conformational adjustments in response to slightly different molecular environments.

Although the cellobiose map shows two 'valleys' or zones of low conformational energy (Fig. 3.2), we could have anticipated the result that emerged from experiment, that one of these is more populated. This zone

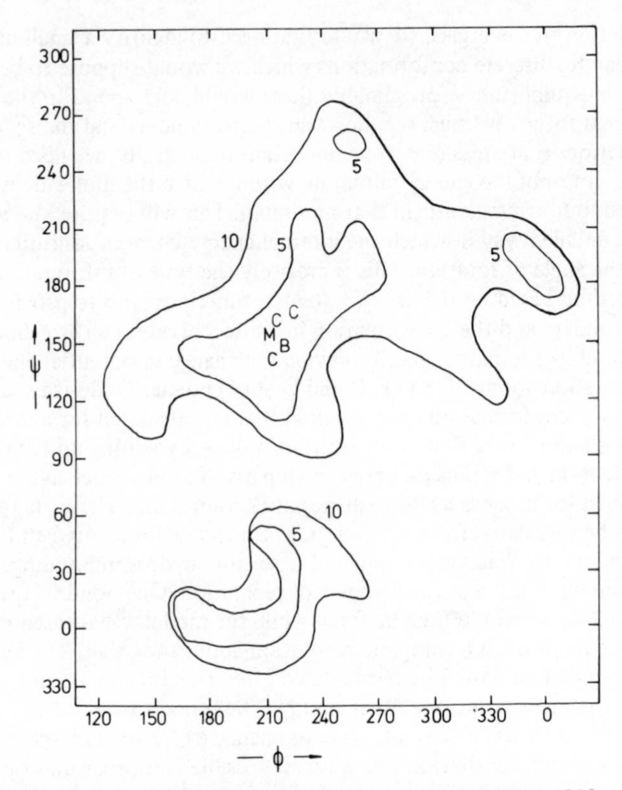

Fig. 3.2 Calculated conformational energy map for cellobiose [8]. Marked on this map by 'C' are the ϕ and ψ values which correspond to conformations found experimentally by X-ray diffraction in crystals of (upper) β-cellobiose, (middle) α-lactose, and (lower) methyl β-cellobioside. Also marked are the conformation of cellulose ('B') and the conformation which is predicted by calculation to correspond to the lowest energy ('M'). Contours are marked corresponding to 5 and 10 Kcal mol^{-1}.

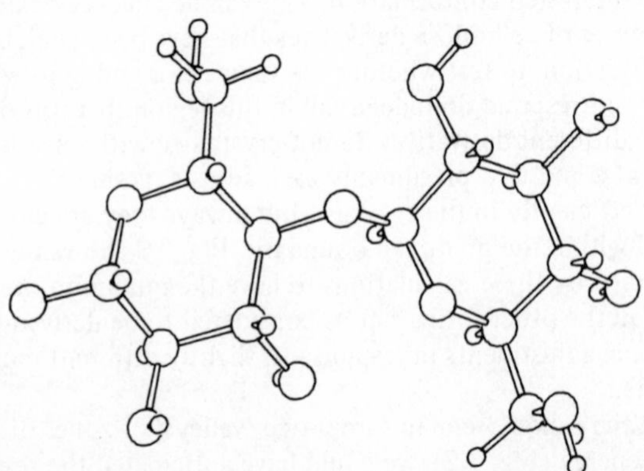

Fig. 3.3 Computer drawing of the forms of cellobiose having minimum energy; see also Fig. 3.5.

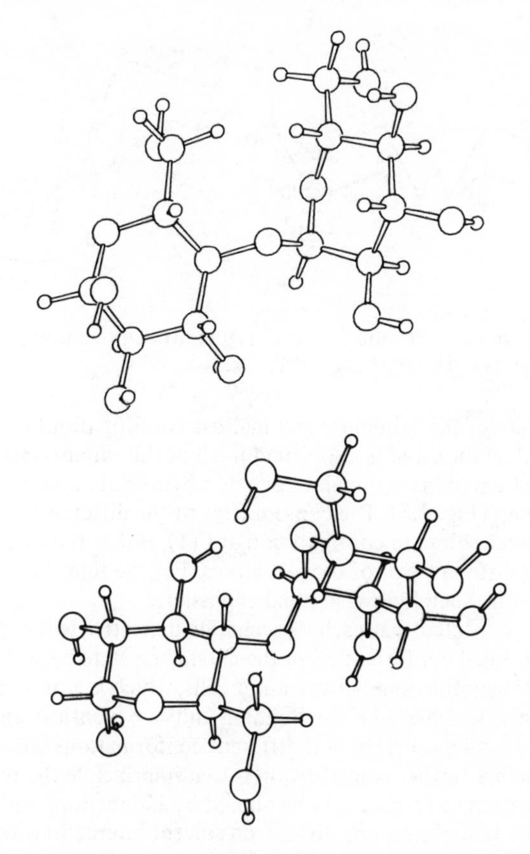

Fig. 3.4 Computer drawings of the forms of maltose which correspond to the minimum energy: upper – within major zone of low energy on the map and: lower – within the minor zone of low energy.

is favoured both because it contains somewhat lower energy states and because it is larger in area and hence more 'probable' for the same reason that the gauche form of *n*-butane is more 'probable' than the anti (Section 1.3). There is no experimental evidence that the β-1,4 linkage ever exists substantially in the minor zone either in the solid state or in solution.

Analogous calculations for maltose again lead to a map with one important and one subsidiary zone of low energy. The form predicted to have the minimum energy is shown in Fig. 3.4(a) and again it is found that conformations characterized by X-ray diffraction for crystalline derivatives are somewhat scattered but mostly clustered in the major zone around this. Here, there is one exception amongst the dozen or more examples, which falls within the subsidiary rather than the major minimum. This is a particular glycoside (6′-iodophenyl α-maltoside) which presumably cannot pack efficiently with the usual type of linkage conformation and therefore adopts the alternative, close to the form shown in Fig. 3.4(b).

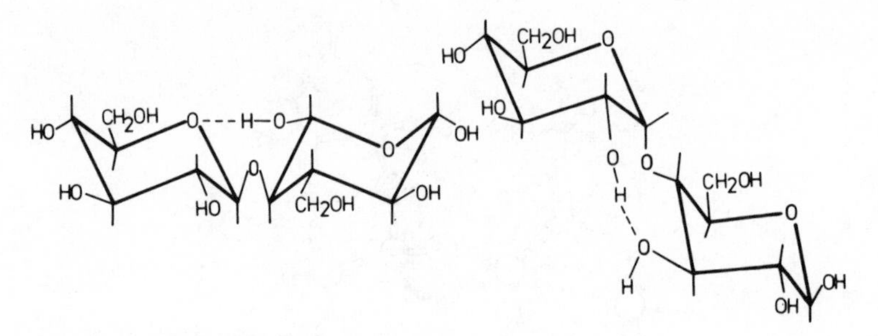

Fig. 3.5 Cellobiose (left) and maltose (right), drawn to show the usual hydrogen bonds between sugar units.

One feature of the cellobiose and maltose conformations that is predicted by calculation and is also found in all of the variants characterized experimentally within each major zone, is a hydrogen bond between the two sugar rings (Fig. 3.5). The consequence of the difference between maltose and cellobiose in configuration at C(1), is that the hydrogen bond joins a different pair of oxygen atoms. For the minor zone of the maltose map, no such hydrogen bond is possible.

It is to be expected that each disaccharide derivative will exist in a single well-defined conformation in the crystalline state because the regular packing within and between unit cells, which is a property of this state, can only be achieved if the building units are identical. In solution, however, it would be surprising if different conformations did not equilibrate and a further complication is that in principle the relative energies of different forms could be altered by interactions with solvent. (Compare the remarks on pp. 26–27, on solvent interactions with monosaccharides.) Methods available for characterizing disaccharide shapes in water solution are much less satisfactory than the beautiful and intricate method of X-ray diffraction which can be applied to the crystalline state. However, it does seem clear from certain interpretations of optical rotation behaviour that cellobiose in water solution stays close to its crystal conformation, no doubt with some fluctuation by bond oscillations and rotations. In contrast, maltose partially populates the minor zone of its energy map, probably because the water molecules compete with sugar hydroxy groups for hydrogen bonding sites to neutralize the energy advantage that conformations in the major zone would otherwise have [9]. It is consistent with this theory that maltose units in nonaqueous solvents exist predominantly in conformations in the major zone [10].

3.3 Chain conformation: order versus disorder

The extra mobility which is a property of linkage conformations in solution compared with the crystal, has particularly important implications for longer chains of sugar units. In a sequence of β-1,4 linked glucose units for example, the shape would fluctuate by independent oscillations between each pair of neighbouring units (Fig. 3.6) and in

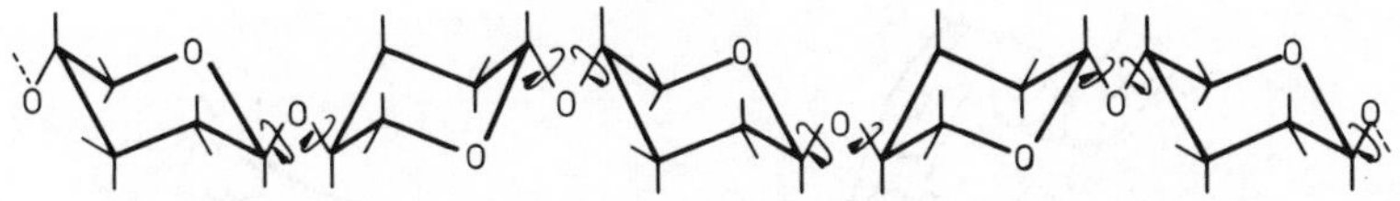

Fig. 3.6 Chain of β-1,4 linked D-glucopyranose units, showing bond oscillations that are expected to occur continuously in solution.

addition individual sugar rings would occasionally flip to higher energy chair, boat, or twist-boat conformations. As a result, a snapshot at any instant (if this were possible) would be unlikely to show the same conformational relationship between each sugar unit and the next along the chain. An instant later, conformational relationships would be further changed. This form of the chain which is fluctuating continuously between different local and overall conformations is known as the *random coil.*

It is also possible to imagine that sometimes interaction energies between sugar units would be sufficiently strong and numerous to fix the conformational relationship between each sugar unit and the next to withstand the randomizing effect of thermal collisions and hence to retain a unique shape. For a chain with a regular sequence of identical sugar residues, it is likely that this unique shape would have the same conformational relationship between successive units because any particularly favourable interactions between one unit and its neighbour would also be possible for the neighbour and next neighbour. This means that the periodic sequence would generate a periodic shape. How this might happen is illustrated by the schematic chain in Fig. 3.7. When this chain, initially fluctuating and disordered, is arranged to form one turn of helix, only one hydrogen bond is possible; on the other hand, three turns allow thirteen hydrogen bonds (Fig. 3.7). Thus, by fixing bond rotations to form 1, 2, 3, 4 . . . . n consecutive turns of this particular helix it is possible form 1, 7, 13, 19 . . . . $6n-5$ hydrogen bonds respectively. The longer the helix grows the more stable it becomes – which means it is more able to withstand thermal collisions. Real helices are stabilised by other forces as well as hydrogen bonding, but the general principle remains that the cooperation of many local forces of attraction to achieve long range order is much more possible than the ordering of particular pairs of consecutive units. These long range influences are known as *cooperative interactions* and are necessary for all biopolymers, including carbohydrate chains, to adopt ordered shapes in solution. The 'all-or-none' quality of the stabilization gives to biopolymer conformations a characteristic property of forming and unfolding sharply rather than gradually.

In the same sense as we can regard the ordered conformational state of a typical biopolymer as stabilized by cooperative interactions, we can say that the disordered state, or random coil, is stabilised by *conformational entropy.* This is because the greater the number of local conformations between which the chain can fluctuate to take up alternative overall shapes, the more difficult will it be to overcome thermal motion to settle

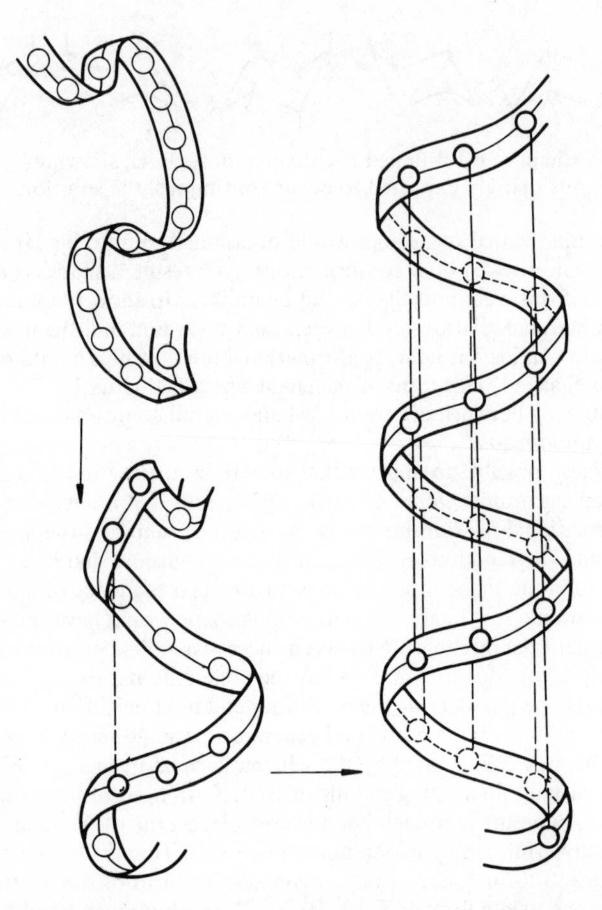

Fig. 3.7 Carbohydrate chain represented schematically as a converting from random coil to helix, through intermediates in which one turn and then three turns of helix have formed. The hydrogen bonds are shown by dotted lines.

in any one particular form. In this way the internal flexibility can operate successfully against forces of attraction which favour a unique polymer conformation. Entropy then offsets potential energy, just as it does when, say, we find a higher proportion of the gauche form of *n*-butane in the equilibrium population that we might at first expect on the basis of its internal energy relative to the anti form (Section 1.3).

4 Simple carbohydrate chains of the periodic type

This chapter will outline the ordered shapes that are possible for carbohydrate chains having the periodic type of sequence, and the extent to which their properties can be predicted or at least rationalized. This will require an examination of the geometry of some of these ordered shapes as they have been characterized experimentally.

The detailed and unambiguous characterization of polymer shapes is possible by only one method, namely X-ray diffraction, and this is applicable only to samples in the solid state which are preferably crystalline or at least have some degree of regularity in the way the molecules in them are packed. To apply this to carbohydrate chains, it might sometimes be necessary to isolate the chains from their biological environment and put them into this condensed form in which they might not naturally exist. The additional question must then be asked whether the shapes characterized in this way have any relevance to the biological form and function, and further experiments can usually be devised to provide an answer. In this way, it is possible to use the X-ray conformation to formulate hypotheses about chain shapes in the biological environment and then to test and refine the hypotheses using other methods. We usually find that whenever carbohydrate polymers are ordered in highly hydrated environments, the ordered form does correspond to a state that has been characterized by X-ray diffraction. However, it must never be taken for granted without proof that the shape will actually be ordered rather than disordered.

4.1 Conformational families

As we explained in Chapter 3, the present methods for predicting carbohydrate conformations cannot give absolute accuracy in detail because they take into account only the attractions and repulsions within an isolated molecule; interactions with surrounding molecules are ignored. Further uncertainties arise because assumptions have to be made about energy functions and such geometrical parameters as bond lengths and perhaps bond angles; these assumptions might not be absolutely accurate down to the last detail. Nevertheless, while admitting that the methods are imperfect, very useful predictions are still possible even though they may only be rather general in nature.

Consider first the regular conformations that are possible for carbohydrate chains containing a single type of sugar unit in a single type of linkage. We know from Section 3.3 that, if and when these periodic sequences adopt ordered shapes, the shapes will themselves be regular and periodic. From the principles of geometry it can be shown that this

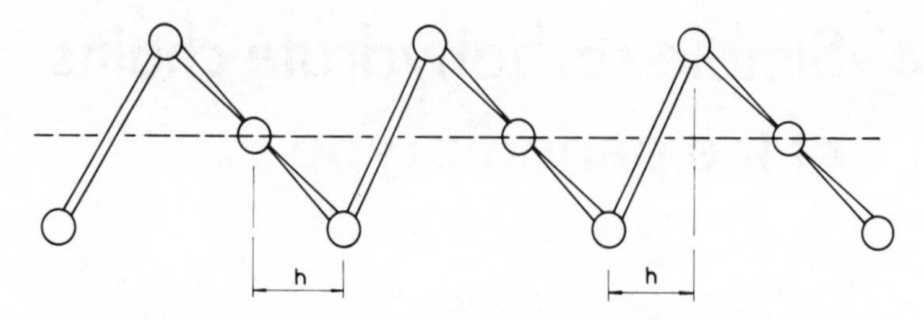

Fig. 4.1 Chain of atoms arranged in a helix for which $n = -3$ and h corresponds to the distance shown.

type of regular shape can always be described as a helix (even though it may not necessarily look like the conventional idea of a helix) and that its general contour can be specified by two parameters; these are n, the number of monomer units per turn of helix, and h, the projected length of each monomer unit on the helix axis, as shown in Fig. 4.1. By convention, a positive value of n is assigned when the helix has the sense of a right handed screw, and a negative value when the sense is left handed. It is often useful to use these parameters instead of ϕ and ψ to define chain conformation because their values are usually deduced very easily from an X-ray fibre diffraction diagram, and they give a good idea of the general overall shape – for example, whether it is stretched out and slim or compressed and squat. If ϕ and ψ are known, it is a simple mathematical task to calculate n and h but the reverse calculation usually gives an ambiguous answer. This is because a polymer chain can usually be arranged in several different ways to follow the same helix contour, from which it must follow that one pair of values for n and h may be fitted by several alternative pairs of values of ϕ and ψ. In the derivation of ϕ and ψ by X-ray diffraction methods, the knowledge of n and h can therefore take us only to the stage of listing a series of alternative answers and further evidence is required to decide which of these corresponds to physical reality.

It is useful to extend the calculations which predict the ranges of ϕ and ψ within which the conformations of a particular glycosidic linkage will occur (e.g. Fig. 3.2), to find the corresponding ranges of n and h. This will tell us the type of overall shape that the chain will adopt if it ever exists in an ordered, periodic conformation. Because of the limitations in the calculations (see above) the result cannot be precise but it does turn out that distinct ranges of n and h are predicted for different homopolymers, depending on the covalent structure. That these correspond to fundamentally different types of chain shape is immediately apparent from representative plots such as for the polymers of β-D-glucopyranose shown in Fig. 4.2. It turns out [11] that four chain families are distinguished as follows:–

Ribbon family, for which n ranges from 2 to $\pm$ 4 and h is close to the actual length of a sugar unit. This means that the unit cannot lie at a large angle of tilt to the helix axis but must be nearly parallel to it.

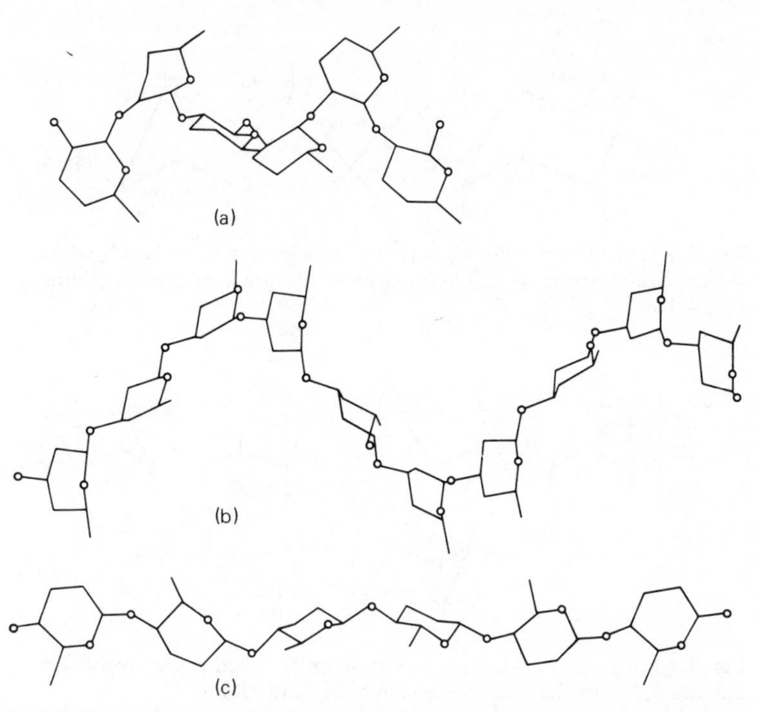

Fig. 4.2 Conformations [11] representative of the range that is energetically likely for:
(a) 1,2 linked polymer of β-D-glucopyranose. (Range predicted was: $n = 4 \rightarrow -2$, $h = 2 \rightarrow 3$Å. Example plotted has $n = -2.62$, $h = 2.79$Å.)
(b) 1,3 linked polymer of β-D-glucopyranose. (Range predicted was: $n = +15 \rightarrow +2$, $h = 0 \rightarrow 5$Å. Example plotted has $n = 5.64$, $h = 3.16$Å.)
(c) 1,4 linked polymer of β-D-glucopyranose. (Range predicted was: $n = -4 \rightarrow +3$, $h = 4 \rightarrow 6$Å. Example plotted has $n = -2.55$, $h = 5.13$Å.)

This implies a pulled out ribbon-like shape. One example is 1,4 linked β-D-glucopyranose (Fig. 4.2).

Hollow helix family for which n can adopt a much wider range of values (e.g. 2 to ± 10) and, even more diagnostically, h can approach zero. This implies a shape like a flexible wire spring that can exist in various states of extension. One example is 1,3 linked β-D-glucopyranose (Fig. 4.2).

Crumpled family, for which the characteristic feature is that many linkage conformations that correspond to a low energy for an isolated disaccharide unit, if repeated periodically, would cause clashes between nonconsecutive units. This means that ordered conformations are more difficult for chains of this type. These are rare in biological systems and will not be further discussed, but one example is 1,2 linked β-D-glucopyranose (Fig. 4.2).

Loosely jointed family, in which the sugar rings are separated by three bonds rather than two as in 1,6 linked β-D-glucopyranose (Fig. 4.3). This leads to extra freedom of rotation not only because of the extra

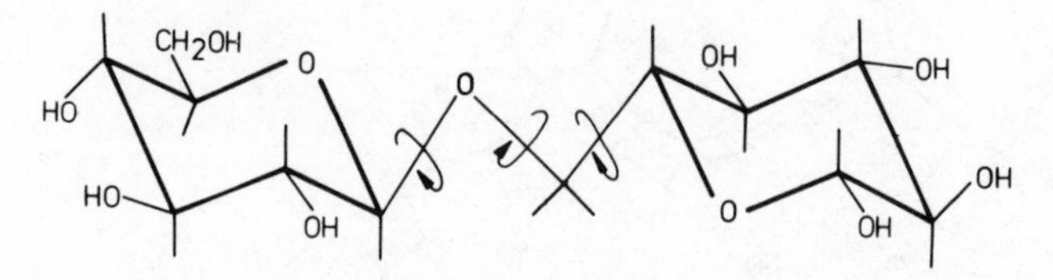

Fig. 4.3 1,6 linked β-D-glucopyranose, showing the three bonds which separate the sugar rings and about which rotation is possible. (Compare Fig. 4.4.)

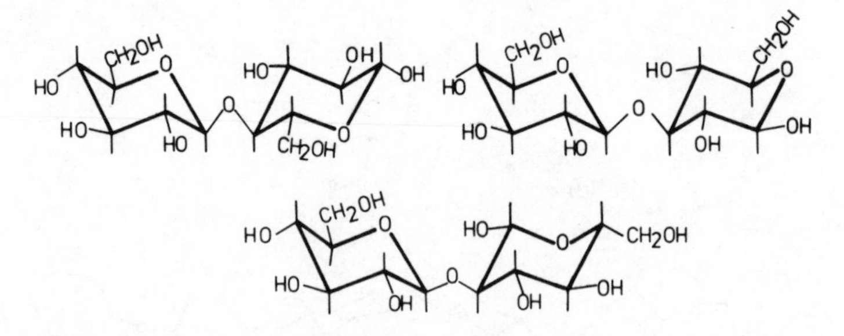

Fig. 4.4 Formulae of β-D-glucopyranose units linked respectively: 1,4; 1,3 and 1,2 – to show the stereochemical similarities.

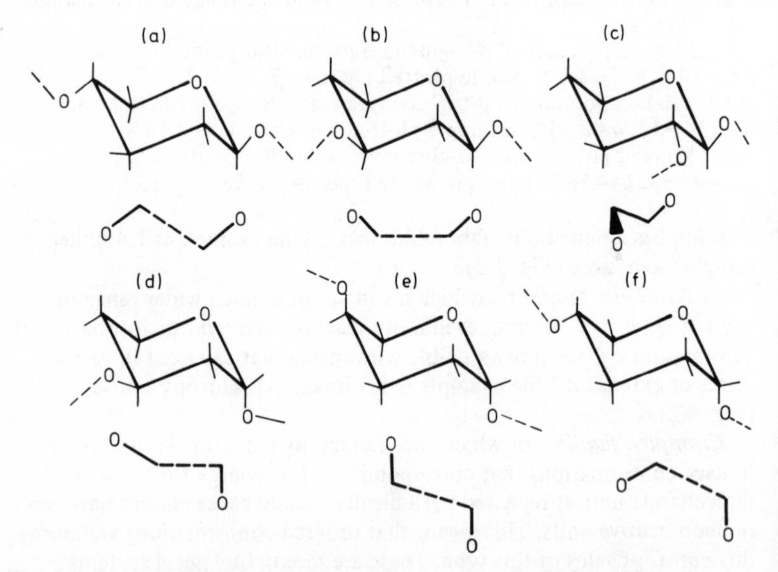

Fig. 4.5 Geometrical relationships across sugar units which determine the conformational family: (a) zig-zag relationship across 1,4-linked β-D-glucopyranose; (b) U turn across 1,3-linked β-D-glucopyranose; (c) twisted form across 1,2-linked β-D-glucopyranose; (d) zig-zag across 1,3-linked α-D-glucopyranose; and (e) 1,4-linked α-D-galactopyranose; (f) U-turn across 1,4-linked α-D-glucopyranose.

bond about which it can occur but also because the sugar rings are further apart and less likely to clash.

The conformation type is not primarily determined by interaction energies between neighbouring sugar units in the chain. This is intuitively obvious from inspection of the conformational formulae for the 1,4, 1,3 and 1,2 linkages in β-D-glucopyranose polymers which lead respectively to chains of the ribbon, hollow helix and crumpled families: the inter-residue contacts across these linkages are so similar (Fig. 4.4) that we would expect the low energy ranges of ϕ and ψ to be quite similar, and this is confirmed by calculations. The primary influence that determines the type of overall chain shape is therefore the geometrical relationships *within* each sugar unit (Fig. 4.5) rather than the interaction energies between them. For these three examples, the significant differences are:–

For the β-1,4 linkage, the bonds from the sugar unit to its two bridging oxygens define a zig-zag form (Fig. 4.5(a)).
For the β-1,3 linkage they define a U-turn (Fig. 4.5(b)).
For the β-1,2 linkage they define a twisted form (Fig. 4.5(c)).

Whenever the zig-zag arrangement exists – other examples are in the α-1,3 linked units shown in Fig. 4.5(d) and the α-1,4 linked galactose units in (e) – the chain is almost always of the ribbon family. When the U-turn arrangement exists – another example is in the chains of α-1,4 linked glucose units in Fig. 4.5(f) – the chain shape is of the hollow helix family.

4.2 Occurrence, properties and function of the ribbon family

4.2.1 Extracellular carbohydrate composites

Carbohydrate-rich structures seem to exist externally to the plasma membrane of most types of living cell but their composition and organization varies widely. For example, they are different in plants compared with animals, in yeasts compared with bacteria, or even in certain algae compared with 'typical' higher plants.

In mature tissue of higher plants, there is a wall structure which depends for integrity on the packing and binding within bundles of carbohydrate chains of the ribbon type. This is an elaborate structure in which the chains show a number of variations on the ribbon theme, and it makes an instructive study in carbohydrate shapes. The bacterial cell wall is quite different; integrity depends on peptide cross linkages and the carbohydrate chains, although members of the ribbon family, may not actually exist in such an ordered state. External to this peptidoglycan framework are other carbohydrate structures, in which the sequences show enormous variety. In some yeasts, interactions for maintaining shape and cohesion within the wall are between carbohydrate chains of the hollow helix type. Around the surfaces of animal cells, however, the external carbohydrate structure is much less rigid and appears to be a loosely organized complex of relatively short carbohydrate chains anchored to protein or lipid components of the cell membrane which in some cases are linked to cytoplasmic structures beneath. We shall first discuss the wall structure in plants because the shapes and interactions of the carbohydrates are much better understood.

4.2.2 The plant cell wall [12]

The wall structure develops to match the function of the cell at each stage of plate life. For example, the fibre cells of wood (Fig. 4.6) have a supporting function and hence their walls are very thick, with a characteristic densely packed arrangement of cellulose microfibrils in several distinct layers; some layers are made more rigid by addition of lignin, a non-carbohydrate polymer with a cementing function (the monomer units are substituted phenylpropanes). These walls also contain other chains of the ribbon family such as polymers of xylose, mannose and galacturonic acid (Fig. 4.6). This dense wall structure contrasts with the wall structure in a young and actively growing tissue, like the root or shoot of a seedling, which is thin and pliable with a high water content and a loose and disordered arrangement of microfibrils to allow for the cell expansion which is necessary because growth in plants involves very considerable cell enlargement as well as cell division. This enlargement phase of cell growth may resemble the inflation of a sausage-shaped balloon, so that the shape becomes longer in the direction of tissue growth. This process, known as *extension growth*, involves the stretching of the side walls and hence alignment of the microfibrils.

Simultaneously with enlargement of the young cell, more cellulose microfibrils are deposited on the wall surface next to the cytoplasmic membrane. These are laid down at right-angles to the direction of growth, so that each layer becomes buried by new layers deposited on top and the microfibrils are continuously reoriented by cell extension in the way that the contour of a wire spring changes when it is pulled out. The cage of microfibrils around the cell is enlarged most easily by extension, just as a wire spring is more easily stretched than dilated. Other polysaccharides, known as matrix polysaccharides, are added at the same time to fill in the gaps in the network of cellulose microfibrils. Eventually, a complex multilayered structure like that of the wood fibre cell (Fig. 4.6) may be built up in this way. The biosynthetic machinery is not fully understood but it must be very elaborate. It seems that matrix carbohydrate chains are formed by polymerisation from sugar nucleotides (p. 30) and exported in Golgi vesicles. The cellulose chains in each microfibril are probably synthesised together by an assembly of enzyme complexes located on the cytoplasmic membrane, and hence extruded in organized form into the cell wall with their alignment and pattern of deposition determined by the cytoplasmic fibrillar protein system known as the *microtubules* [13]. Higher levels of control must exist to direct the overall organization but the mechanisms are quite unknown at present.

In summary, the mechanical properties of a plant cell wall are crucial

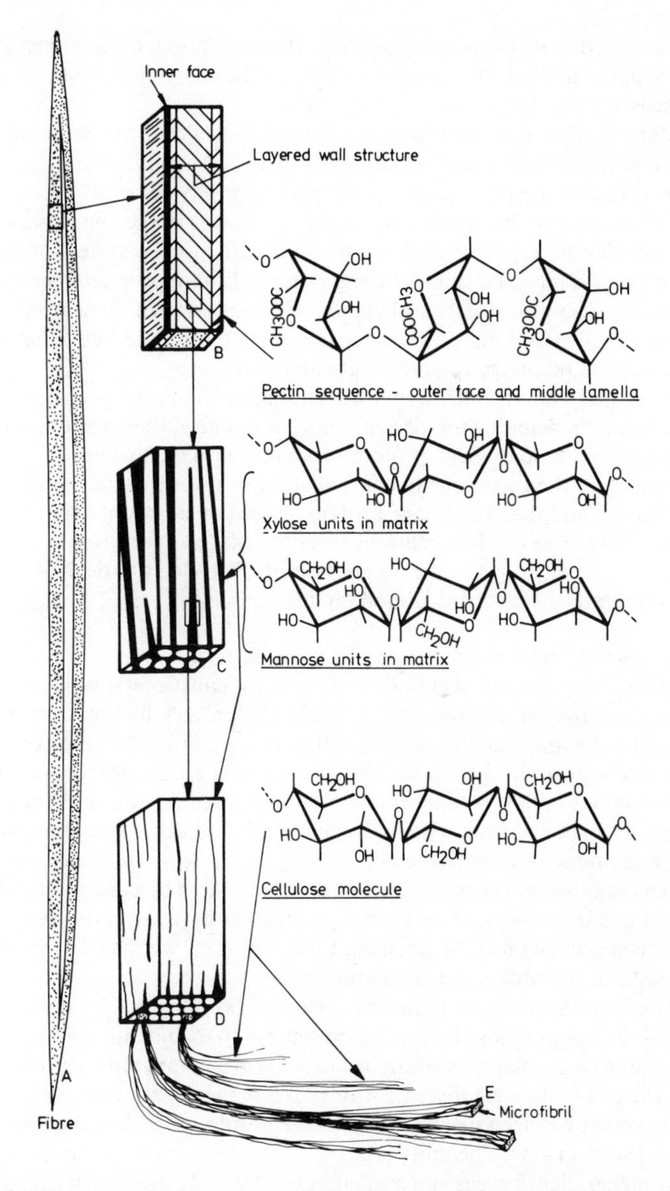

Fig. 4.6 Representation of cell wall structure in a fibre cell from wood. Fibre (A) has a layered structure (B). In a fragment of the central layer of this wall (C), deposits of cellulose (white) are embedded in a matrix (black), of other polysaccharides and lignin. The cellulose deposits consist of many microfibrils (white in D) further embedded in matrix. Microfibrils consist of bundles of cellulose molecules in crystalline packing (E). Some polysaccharide components of different wall regions are shown; note that these are partial structures only (see Chapter 5). This diagram is adopted from [12].

to the function of the tissue – whether this is to provide the strength and support in the trunk of a forest tree, or the controlled plasticity that is necessary for the growth of young tissue.

Many further examples could be given of specialized wall structures which perform their function through a form and combination of properties that are appropriate to the particular physiological function. In the ripening of fruits there are dramatic changes in the consistency of intercellular material to soften the tissue. In the vascular system of plants there are continuous channels having rigid walls of appropriate permeability. The wall structures in seeds are tough enough for survival during the dispersal stage but responsive enough for rapid hydration and mobilization of energy reserves in germination.

In all these intercellular structures and many more, the packing of carbohydrate chains of the ribbon family is a central theme in overall organization. Biopolymers of other types may also be involved – examples are the final stage in consolidation of woody tissue by lignin deposition, and the presence of protein or peptide components in many walls. Every type of plant cell wall also depends, however, for its integrity on controlled non-covalent bonding between a variety of carbohydrate chains of the ribbon family.

4.2.3 Carbohydrate sequences in plant cell walls

Even when a plant cell wall polysaccharide contains several types of sugar unit, sugar linkage, and sequence of units and linkage, its function can be understood by considering the properties of each individual type of sequence separately. The most fundamental of these sequences are those of the ribbon family because they function in bonding, through their ability to align and pack with hydrogen bonds and other favourable non-covalent interactions between chains. In Chapter 5, we show that the sequences of other types can modulate this bonding by controlling its extent and its pattern, or by forming interstices in which water and other components may be retained. First, however, we shall discuss in this section the ribbon sequences themselves.

The building units for these sequences are (a) β-D-glucopyranose (b) β-D-mannopyranose (c) β-D-xylopyranose (d) β-D-mannopyranosyluronate (e) α-D-galactopyranosyluronate (f) α-L-gulopyranosyluronate (see Chart 1). They are linked 1,4 to generate the zig-zag type of arrangement between neighbouring oxygen bridges, and hence a ribbon shape (Section 4.1). The units (a), (b), (c), and (e) are found in all or most higher plant tissues and a number of lower plants as well; (d) and (f) are peculiar to brown algae (Phaeophyceae) but are included because they illustrate principles which apply more generally. Plots of the detailed shapes of these ribbons as derived by X-ray fibre diffraction analysis, show that they differ in important details (Fig. 4.7). Cellulose and mannan, the polymers of (a) and (b) respectively, are flat and extended; xylan and polymannuronate, polymers of (c) and (d) are twisted; polyguluronate, the polymer of (f) is buckled, and polygalacturonate, the polymer of (e) is both buckled and twisted.

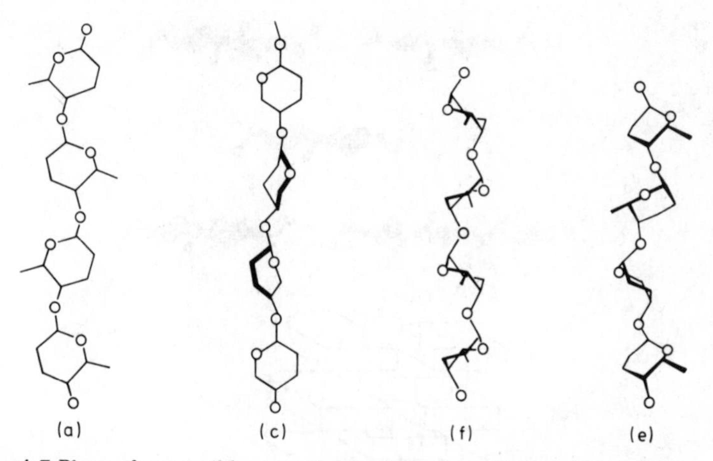

Fig. 4.7 Plots of some ribbon conformations: (a) cellulose [15,16] (the mannan conformation is very similar to this); (c) xylan [17] (the poly-mannuronate conformation is related to this; (f) polyguluronate [18]; (e) polygalaeturonate [19].

Flat ribbons are characteristic of the cell wall components which appear to be most firmly bound because they are most difficult to remove from the organization by solvents such as alkali, and also because X-ray diffraction shows that they have the highest degree of ordered chain packing. This might suggest that the flat ribbons are most easily aligned and bonded. The high strength, fibrous character, insolubility and inertness that are so characteristic of cellulose are understandable for such compact and tightly bonded aggregates. X-ray diffraction shows that the ribbon-like chains in native cellulose are probably [15] parallel rather than anti-parallel – these two possibilities arise because the two ends of a carbohydrate chain are different, one end being unsubstituted at C(4) and the other unsubstituted at C(1). In the parallel mode, all chain ends of a given type occur at the same end of the bundle (Fig. 4.8(b)). The ribbon-like chains lie side by side in sheets, joined by hydrogen bonds. These sheets are laid on top of each other in a way that staggers the ribbons, much in the way that bricks are staggered from one layer to the next in building a wall (Fig. 4.8(a) and (b)).

In industrial processing, cellulose is often treated severely with alkali, as in the so-called mercerization process to make cotton easier to dye, and in the regeneration of cellulose to manufacture rayon or cellophane. The chains then rearrange to a new packing form, known as cellulose II (the native form is termed cellulose I). There is no known way of converting cellulose II back to cellulose I. The chain conformation is essentially the same in both forms but, in regenerated cellulose, the chains are anti-parallel and piled on top of each other with more resemblance to planks in a timberyard than to bricks in a wall (Fig. 4.8).

A further important influence that contributes to the strength of binding in cellulose is the stiffness of the chains. This follows from the

(a)

(b)

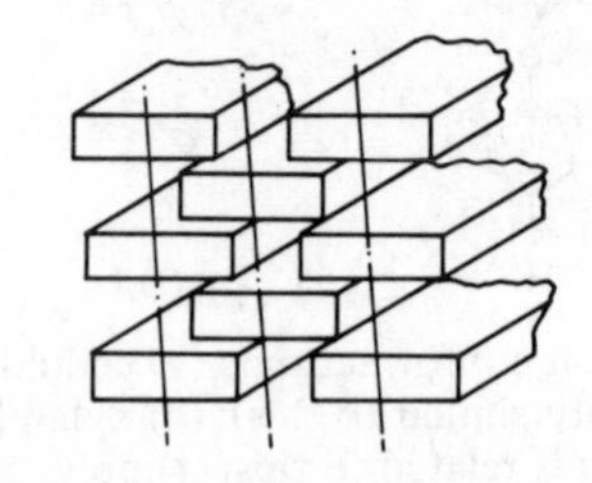

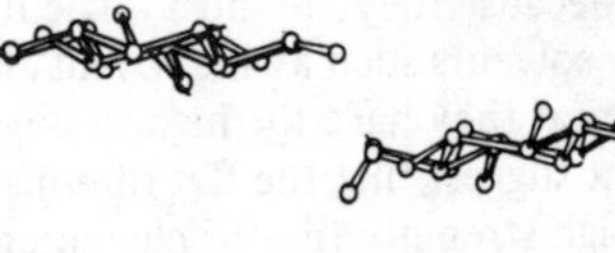

(d)

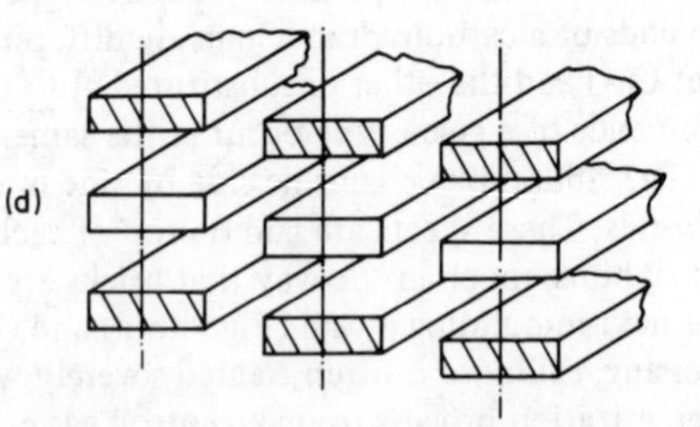

Fig. 4.8 Packing of cellulose chains in the natural and regenerated states: all views are end-on, down a bundle of chains. (a) Computer plot of chain packing in natural cellulose ('Cellulose I'): note that the ribbons are layered in sheets and staggered between sheets, as shown schematically in (b); chains are parallel rather than anti-parallel i.e. the chain ends at one end of a bundle are all of one type (not hatched); (c) Computer plot of chains packing in regenerated cellulose ('Cellulose II'): note that the ribbons are piled in more obvious stacks as shown schematically in (d); chains are anti-parallel i.e. chain ends at the end of a bundle are of both types (hatched and not hatched).

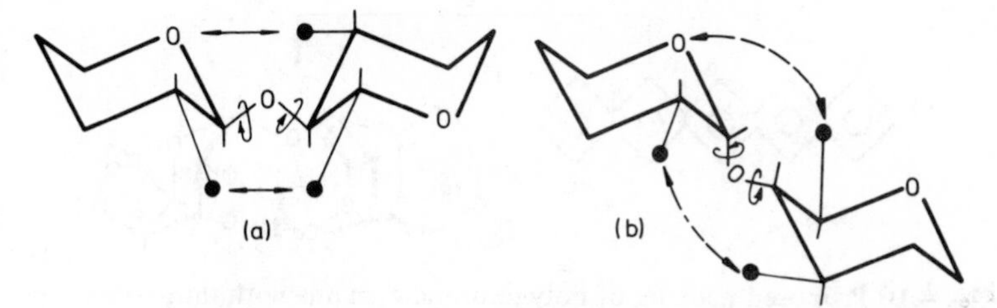

Fig. 4.9 Equatorial groups which have an important influence on the flexibility at linkages between sugar units (●). Flexibility is further restricted if an equatorial bond to the glycosidic oxygen is changed to axial (e.g. as in (b) compared to (a)).

arguments of Section 3.3, in which it was shown that an order → disorder conversion is assisted by conformational entropy i.e. by the flexibility of the disordered state. Carbohydrate chains are not very flexible in comparison with other biopolymers. Conformational energy calculations show that:

(i) Flexibility is particularly restricted when the equatorial groups adjacent to the linkage are large. The reason for this is easily seen from molecular models – the larger these groups are, the sooner they clash when the bonds are rotated away from the minimum energy position – see Fig. 4.9.

(ii) Flexibility is further restricted with increase of the number of axial bonds to the glycosidic oxygen (Fig. 4.9).

The twisted ribbons, such as xylan (Fig. 4.7(c)) are less firmly bound into the cell wall as judged by the criteria of the ease with which they can be extracted and the extent to which they are crystalline in the native state. Evidently they pack less well than the flat cellulose ribbons. The greater flexibility of the xylan chain is no doubt another contributing factor. Mannan is also more flexible than cellulose and is less firmly bound into the wall.

Because of the shape of buckled ribbons, they might be expected to leave interstices when they pack together. The bonding between them would only be strong if such voids could be filled, for example by small species such as water or ions. In fact the buckled form is characteristic of carbohydrate anions such as polyguluronate and polygalacturonate for which a suitable cation can not only fit the cavity but also screen the electrostatic repulsion between like charges that would otherwise cause the chains to repel. One important cation to be involved in such complexation in biological systems is calcium, Ca^{++}. The association is cooperative (see Section 3.3) because binding of the first cation between any pair of chains causes alignment which facilitates binding of the next, and so on along the sequence. The interaction energy for binding a single cation between two chains is of course too weak to hold them together firmly enough to resist thermal collisions at normal temperatures, and it has been shown by experiment that a sequence of about twenty guluronate or galacturonate units is needed. This stability is

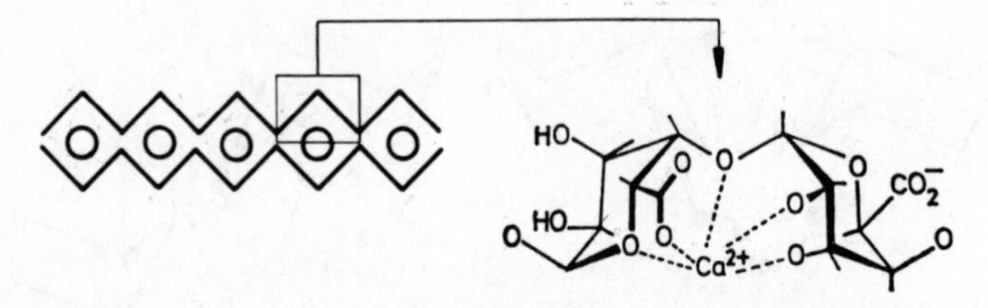

Fig. 4.10 Proposed packing of polyguluronate chains with interstices filled by Ca^{++} ions. Left, schematic representation of two buckled chains (each by a zig-zag line) with Ca^{++} ions represented by circles. Right, detailed conformational representation of a possible mode of cation coordination by one chain.

believed to be enhanced because sufficient oxygen atoms on each chain are suitably placed to form a complete coordination sphere (Fig. 4.10). This form of association has been named the 'egg-box model' because the array of ions is held between the carbohydrate chains in much the way that eggs are held between the trays of an egg-box [20].

It should now be apparent that different carbohydrate ribbons contribute in different ways to bonding in plant cell walls. The stiff, flat cellulose chain is well adapted for its function as the main structural element, whereas the more flexible mannan chain and even more so the flexible and twisted xylan chain, form bonds which are more easily rearranged. The buckled form of polygalacturonate, confers the ability for strong cohesion with the involvement of cations which may be significant for the regulation of bonding and/or ionic relationships. All of these must play a part in the complex series of events in wall development but we have as yet very little understanding of this in detail. In Chapter 5, we shall show that these relationships are even more complicated than we have indicated yet, because each carbohydrate sequence can actually occur as an entity in a larger and more complex molecule in which the characteristic properties are further modulated by the influence of additional features.

4.2.4 Ribbon sequences in other extracellular structures

In crustaceans, insects and spiders, the polysaccharide chitin is a basic skeletal material since it contributes to the structure of the exoskeleton, the lining of the gut, the tendons, wing coverings and the internal skeleton. This polysaccharide is also present in the cell walls of fungi. Perhaps therefore it is even more widespread and abundant in nature than cellulose – it has been estimated that about 10^{10} to 10^{11} tons of new cellulose and chitin are synthesized naturally each year on the earth's surface. The structure of chitin (Fig. 4.11(a)) is very closely related to cellulose and indeed is identical except that the –OH group on each C(2) is formally replaced by $-NHCOCH_3$; the way in which the chains pack side by side in a crystalline, strongly hydrogen bonded manner, is also very similar. As for cellulose, different forms are known [21] – designated α-, β- and γ-chitin – all of which are built up from sheets of

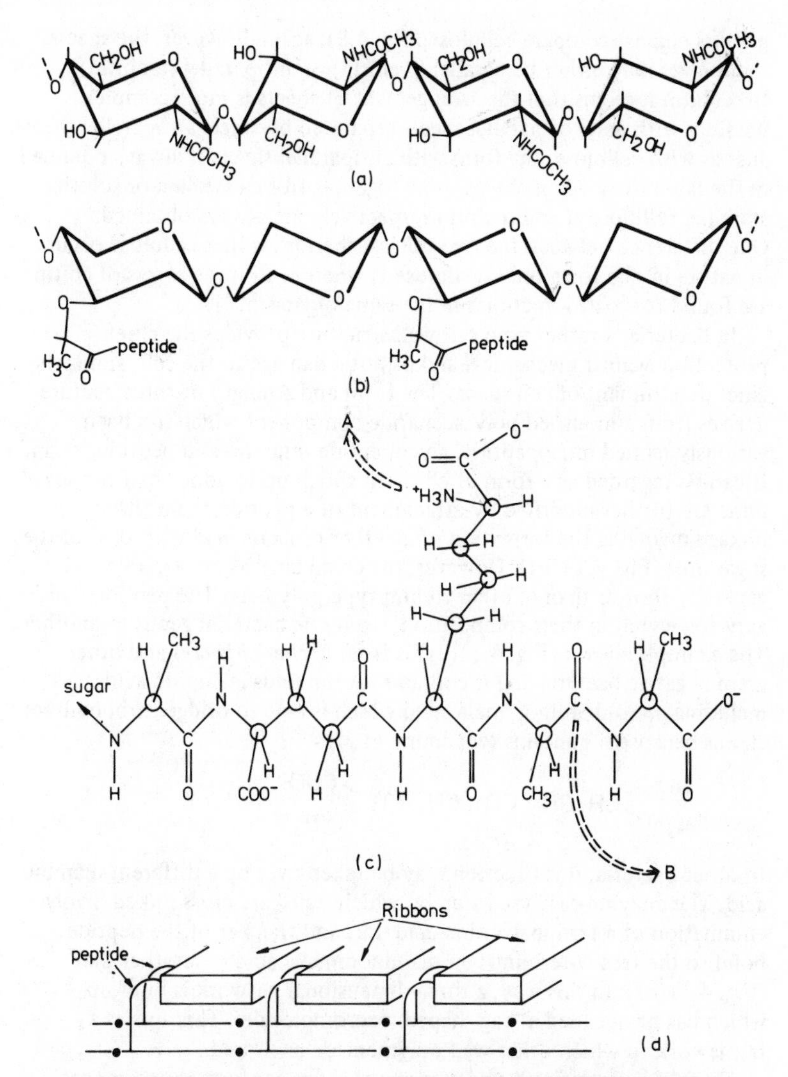

Fig. 4.11 Representation of some carbohydrate chains of the ribbon type: (a) Chitin. (b) Peptidoglycan – the substituents on sugar rings are omitted for clarity (they are the same as for chitin, except for the side chain). The peptide may have different structures in different bacteria but one type is shown in (c) which is L-alanine–D-glutamate–mesodiaminopimelic acid–D-alanine–D-alanine. Cross-links into the wall are by new peptide bonds through the positions shown by the arrows A and B, with loss of terminal D-alanine. (Some carbon atoms are shown as balls to facilitate display of the stereochemistry.) (d) Schematic of a possible mode of cross-linking of peptidoglycan. This is an end-on view of a sheet of carbohydrate ribbons, assumed for the sake of illustration to be parallel rather than anti parallel. Cross-links between peptides are shown as filled circles.

parallel chains (compare cellulose, Fig. 4.8); again, however, the sheets themselves may either be parallel (β-chitin) or antiparallel (α-chitin). In γ-chitin it seems that the arrangement of sheets is more complex, possibly with pairs of parallel sheets separated by single antiparallel sheets. Just as with cellulose, the form with antiparallel sheets is always obtained in the laboratory when chains some together from a swollen or solution state i.e. cellulose II and α-chitin respectively are always obtained. One difference between the two polysaccharides is that cellulose occurs in nature in one form only (cellulose I) whereas all three forms of chitin are found to exist, sometimes in the same organism.

In bacteria, a rather rigid cell wall structure provides the chief protection against mechanical and osmotic damage to the cell, and is the chief determinant of cell shape. The form and strength of this structure derives from a modified polysaccharide component which has been variously named mucopeptide, glycopeptide, murein, and peptidoglycan. It can be regarded as a form of chitin in which up to about half the sugar units are further modified by attachment of a peptide chain through a linkage involving the formation of an ether of lactic acid with 0(3) of the sugar unit (Fig. 4.11(b)). However, the chain lengths are variable and generally shorter than in other ribbon type polymers. The peptide 'tails' vary somewhat in their composition, from one bacterial genus to another. The example shown (Fig. 4.11(c)) is from *Escherichia coli* and other gram negative bacteria and it contains several unusual amino acids including *meso*-diaminopimelic acid which is able to bridge carbohydrate chains because it contains two amino groups:

$$\begin{matrix} ^{-}OOC \\ ^{+}H_3N \end{matrix} \!\!>\! CH.\, CH_2.\, CH_2.\, CH_2.\, CH. \!<\! \begin{matrix} COO^{-} \\ NH_3^{+} \end{matrix}.$$

In other bacteria, this function may be taken over by a different diamino acid. The enzyme-catalysed step by which 'tails' are cross-linked involves elimination of a terminal amino acid unit and transfer of the peptide bond to the free *N*-terminus on diaminopimelic acid of another 'tail' (Fig. 4.11(c)). In this way, a three-dimensional network is built up which has been called a 'bag shaped macromolecule'. This acts as a framework to which other wall polymers are anchored.

There is little definite evidence about chain conformations in peptidoglycans, or whether the interactions of ordered shapes have any importance in determining properties of the wall. A number of possible models have, however, been proposed and discussed in the literature and there is general agreement on some likely features [22]. When the stiff, ribbon-like carbohydrate chains are aligned in sheets like those in cellulose and chitin, and the peptide tails are put into a likely conformation, it is found that each tail can drape round its ribbon to project away at right angles. A grid-like structure is then formed when they connect (Fig. 4.11(d)). In principle, hydrogen bonding and other non-covalent forces could exist between carbohydrate chains in this arrangement, similar to those in the cellulose and chitin sheets. However, X-ray

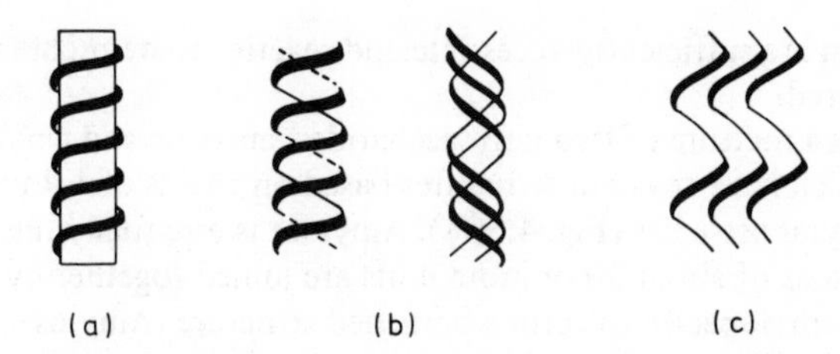

Fig. 4.12 Schematic picture of ways in which hollow helices may fill their voids: (a) inclusion complex; (b) double or triple helix; (c) 'nesting'.

diffraction of isolated wall preparations shows no evidence of ordered packing over long ranges. Although other explanations are possible, perhaps the wall is best regarded as made up of sheets of carbohydrate chains which are disordered to allow some flexibility but reinforced by peptide cross-links to compensate for the loss of strength that must be a consequence of disorder.

4.3 Occurrence, properties and function of the hollow helix family

Two important sugar units which, when present in homopolymeric sequences, give rise to chains of the hollow helix type are shown in Fig. 4.5: these are 1,4 linked α-D-glucopyranose and 1,3 linked β-D-glucopyranose. We have already pointed out that they have the U turn relationship between the bridging oxygen atoms on either side of each sugar unit, rather than the zig-zag relationship which generates a ribbon.

An empty void along the axis of a hollow helix would not be expected to be stable in a natural situation ('nature abhors a vacuum') and, depending on the state of extension of the helix:

(i) When the helix is in its least extended state it can fill the void by formation of an inclusion complex with small molecules which happen to have the appropriate size and shape (Fig. 4.12(a)).

(ii) When the helix is more extended, it may be possible for two or three chains to twist around each other like serpents in an embrace to form a double or triple helix. In such a close-packed structure, each chain can fill the void of the other (Fig. 4.12(b)).

(iii) When the helix is still more extended, the chains can 'nest' and close-pack without twisting round each (Fig. 4.12(c)).

4.3.1 Energy reserve polysaccharides: starch and glycogen

One example of a carbohydrate chain of the hollow helix type is starch which occurs almost as widely in nature as cellulose, although not quite in such large quantities. Although starch and its counterpart in the animal kingdom, glycogen, are not stored as structural materials but as energy reserves, the conformations are relevant to biological function because they allow the molecules to be packed together and accumulated as concentrated deposits. These deposits are structured so that osmotic relationships within the tissues are not disturbed and yet the molecules within them are sufficiently accessible and reactive to be mobilized when required.

Starch is a mixture of two polysaccharides, amylose and amylopectin, both of which have covalent structures based on chains of 1,4 linked α-D-glucopyranose units (Fig. 4.5(f)). Amylose is essentially linear, whereas chains of about 20 or more units are joined together by α-1,6 linkages in amylopectin to form a branched structure. Amylose crystallizes in different forms depending on its environment (Fig. 4.13). In starch granules, the chains are naturally crystalline in forms known as the A form (in cereals) and the B form (in tubers). These are not yet fully characterized but evidence is accumulating that the A form at least is a double helix [23] (Fig. 4.13(a)). This can be converted by appropriate treatment to more extended forms in which the chains presumably pack by 'nesting' and/or by filling voids in the crystal structure with small molecules and ions (Fig. 4.13(b),(c),(d)). Less extended forms can also be obtained, by crystallizing in the presence of suitable guest molecules to form inclusion complexes (Fig. 4.13(e),(f),(g),(h)).

The only form of amylose which has so far been rigorously characterized by X-ray diffraction, is V-amylose. This forms in the presence of suitable small molecules, such as in the familiar complex with iodine. It is responsible for the blue or blue-black colour in the test for starch in plant tissues, and when starch is used as an indicator in iodometric titrations or in testing for oxidizing agents with starch-potassium iodide paper. The chain is coiled into a helix with six glucose units per turn forming a hollow tube (like a wire spring) which encloses a chain of iodine molecules lined up along its axis, as shown diagrammatically in Fig 4.13(f). This helix [24] has a left-handed screw sense and has an efficient system of hydrogen bonding; successive sugar units along the chain are hydrogen bonded between 0(2) and 0(3) as they are in the corresponding disaccharide, maltose (Section 3.2) and additional hydrogen bonds exist between 0(2) and 0(6) of glucose units which are separated by five units in the chain and are therefore adjacent on the helix surface (Fig. 4.13(e)). A further important source of helix stability is no doubt the bonding which holds the enclosed molecule. This must be so, because the tendency to enter the V conformation is greatly enhanced when a suitable substance is present to become complexed, and because this conformation changes slowly to the B form if left in moist air after removal of the enclosed substance. It seems that almost any organic molecule can be complexed provided it is hydrophobic and has a suitable diameter to fit inside. Examples of molecules which can be complexed are phenols, aryl halides, *t*-butanol, cyclohexane and other aliphatic molecules. A reason for this preference for hydrophobic guest molecules is suggested by inspection of the interior of the hollow helix. The D-glucopyranose units are so arranged that the axial C–H bonds on C(3) and C(5) as well as one of the H atoms on each C(6) point inwards to give some hydrophobic character to the lining of the void. The bridge oxygens of each connecting linkage are also part of this lining, and their unshared electron pair orbitals point inwards. We might therefore expect

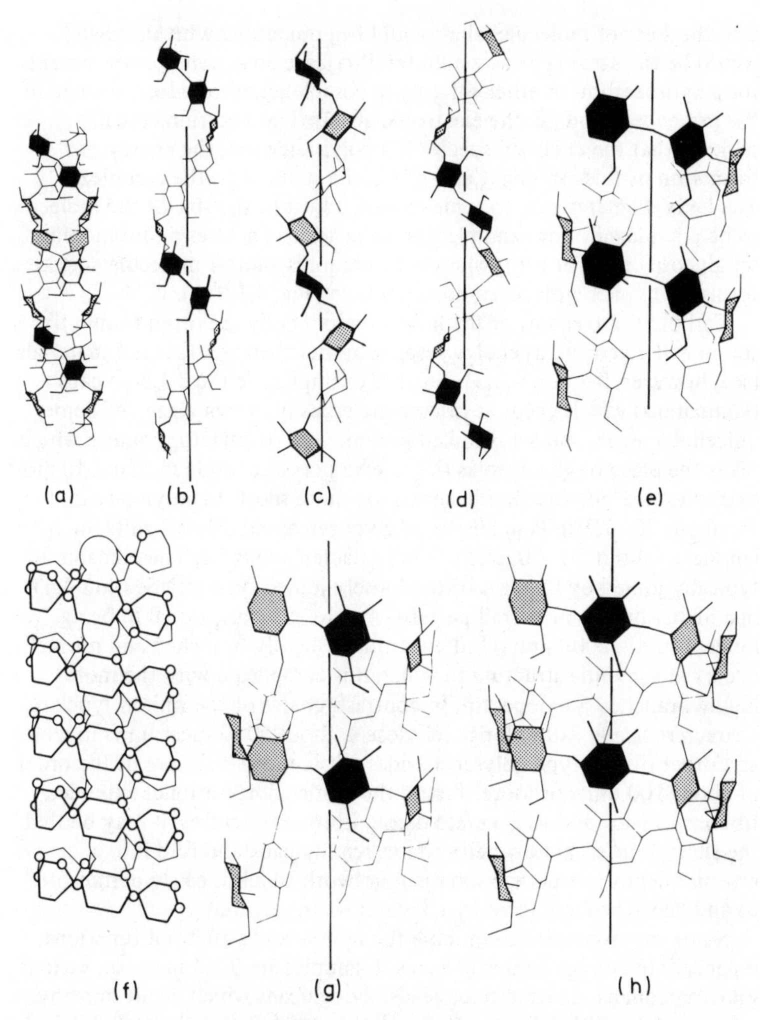

Fig. 4.13 Chain conformations of amylose, all of which can be regarded as the same left handed helix in various states of extension. (a) Double helix proposed for natural amylose in the A form; (b) more extended helix which forms in the presence of potassium hydroxide (six glucose units per turn of helix); (c) still more extended helix which forms when amylose is crystallized with potassium bromide (four units per turn); (d) an intermediate state of extension that is prepared by recrystallization of (b); (e) compressed form of the helix with six units per turn – the so called V form – which is obtained in certain inclusion complexes; (f) diagrammatic representation of inclusion complex of (e); (g) and (h) even more compressed forms with wider channels down the helix axis and eight units per turn, obtained by crystallizing with larger guest molecules. Note that conclusive evidence for the detailed structure is available only for (e) [24]. This diagram is adapted from [25]. The shading of rings is to help perspective in three dimensions.

that the sorts of molecules that would be compatible with the cavity would be the same type as would tend to leave an aqueous environment for a hydrocarbon or ether solvent. In complexes with iodine, formed in the presence of iodide, the electronic absorption spectrum actually suggests that the enclosed species is a polyiodide ion; the energy of formation of this ion might also give some stability to the complex. The helix diameter can, to some extent, adjust to the size of the molecule to be complexed. For example, the helix which encloses *n*- butanol has six glucose units per turn, whereas *t*-butanol is a fatter molecule and so forms a complex with seven units per turn (Fig. 4.13(g)).

Carbohydrate chains of the hollow-helix family are often found, like starch and glycogen, as energy reserves rather than as structural materials (see, however, Section 4.3.3). Further examples are the β-1,3-glucans (laminarans) which occur as reserve materials in brown algae and some unicellular algae, and a 1,3 linked polymer of β-D-galactopyranose which takes the place of glycogen as the reserve polysaccharide in snails. In most examples, the polysaccharide chains are quite short. In amylopectin, they are about 20–25 units in length, in glycogen about 10–12 units, in laminaran about 20–30, and in snail galactan about 10. These chains are typically joined by linkages of the loosely-joined type (see Section 4.4) in a highly branched overall pattern. Amylose is exceptional in being longer (1000–2000 units) but even this is slightly branched and normally occurs in a granule structure in which it is complexed with the more highly branched amylopectin. In contrast, chains of the ribbon type are characteristically rather long: cellulose is about 5000 sugar units in length and other ribbon type polysaccharides of plant cell walls are of the order of about 100 units or more. Just as the tendency is for long, extended ribbons to pack and to generate dense, fibrous structures, it may be that the short, jointed hollow helix chains readily pack and fold into a granular deposit or perhaps an open network which is easily permeable to and hence broken down by enzymes when required.

Some polysaccharides combine the reserve and structural functions, especially in storage organs of seeds. Examples are β-1,4 mannan, various galactomannans in leguminous seeds, xyloglucans which occur in many other seeds, and the glucans of cereal grains which contain both β-1,3 and β-1,4 linkages. Unlike the reserve polysaccharides of the hollow helix family these do not exist in granular deposits but in the form of thickened cell wall deposits in organs that have been specially adapted for long-term storage during dormancy. Consistently with this, their chains are of the ribbon family.

4.3.2 Participation of hollow helices in fatty acid biosynthesis [26,27]

There is good evidence that inclusion complexes formed by hollow helix chains could be important in the control of transport and biosynthesis of fatty acids in some bacteria. In *Mycobacterium smegmatis,* polymers of glucose and of mannose [28] seem to be involved in this way. The glucose polymer is the better characterized. It is a chain of seventeen 1,4 linked α-D-glucose units which carries a side chain of a single sugar unit and is terminated at C(1) by D-glyceric acid; it also has *O*-methyl, *O*-succinyl and other *O*-acyl substituents. Several different forms exist which appear to differ only in the levels of substituents. In the mannose polymer also, the main linkage is α-1,4 and most residues are adjacent monomethyl ethers; no other substituents have yet been identified in this structure.

Synthesis of fatty acids by the enzyme systems of *Mycobacterium smegmatis* proceeds from the usual substrates, namely acetyl-CoA and malonyl-CoA; the first stage is growth of the acyl chain to a length of 16 or 18 carbon atoms, after which the chain is either discharged from the enzyme by the action of palmityl transacylase to form palmityl-CoA, or as enzyme-bound palmitate it may be elongated to as many as 24 carbon atoms. Each of the two polysaccharides can influence the course of synthesis *in vitro*, in a number of ways including (a) synthesis of palmitate is stimulated, (b) the mixture of fatty acid products is shifted in composition towards products of shorter chain length. These effects have been traced [26] to an ability of each polysaccharide to form a 1:1 complex with palmityl and other longer-chain acyl-CoA derivatives and hence to perturb the equilibria that would otherwise exist between various bound forms of the fatty acids. An obvious mechanism to consider for the complexation would be the formation of an inclusion complex analogous to Fig. 4.13(f). Indeed, when the chain of ten 6-*O*-methyl-D-glucose or 3-*O*-methyl-D-mannose residues is arranged in this way, it is seen immediately that the methyl groups project inwards to enhance the hydrophobic character of the channel. The helix matches the palmitate chain closely in length and diameter and therefore an inclusion complex seems very plausible, stabilized by hydrophobic interactions. Possibly this functions in the regulation of feedback inhibition of fatty acid synthesis and in the control of the chain-length distribution of product. The fact that the polysaccharides provide a 'hydrophilic overcoat' for the fatty acid chains would also suggest a role in transport through aqueous environments.

4.3.3 Hollow helices with structural functions

Although the carbohydrate chains which function as extracellular framework materials are usually members of the ribbon family, there are a few examples of hollow helices sufficiently extended to twist around each other to build up a strong network which performs the function in a different way. These may be no more than curiosities, or relics of experimentation by the evolutionary process, because they occur only in a few organisms which may not be wholly representative of their Phyla. One example is the 1,3 linked polymer of β-D-glucose which seems to serve as the structural component of the cell walls of certain yeasts and, together with cellulose, in the pollen tubes of some plants. When isolated, these chains can be crystallized as triple helices [29] (Fig. 4.14). The geometry is such that O(2) of each unit projects into the centre of the helix to hydrogen bond with corresponding atoms on the other two chains (Fig. 4.14(b)). Such twisting together of polysaccharide molecules

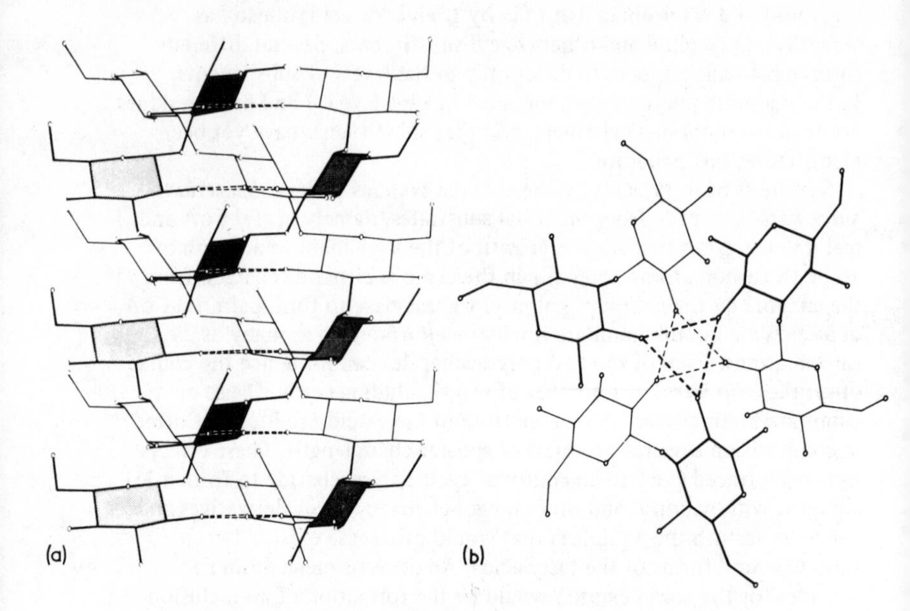

Fig. 4.14 Triple helix of 1,3 linked β-D-glucopyranose: (a) side view; (b) view down the helix axis of the three units which hydrogen bond to help hold the strands together. Shading of rings is to help perspective in three dimensions. (Diagrams provided by Dr. A. Sarko.)

cross-links the chains in units of three rather than in the larger assemblies formed by packing of ribbon-like polysaccharides. This alternative mechanism for building biological cohesion might perhaps create a framework which is more flexible, open and less dense, similar in these respects to the bacterial cell wall (Section 4.2.4).

Another example of cross-linkage by helix formation to provide a basis for a cell-wall skeleton is in certain green algae where 1,3-linked β-D-xylose chains exist as a similar triple helix which had been characterized earlier [30].

4.4 Loosely jointed linkages and chains

The most common linkages of the loosely jointed type are 1,6 between pyranose residues. These are widespread in nature [31], especially in branched polymers, and indeed there are few examples of multi-chain polysaccharides which do not have 1,6 linkages in their structure. The characteristic freedom of rotation about these linkages can provide a bush-like molecule with flexible joints to facilitate biological interactions such as access of enzymes to the interior. Examples are amylopectin, glycogen, and various plant exudate gums. In animal systems, such multi-chain structures may have the carbohydrate chains attached to a polypeptide backbone (Chapter 5) through linkages which are stereochemically

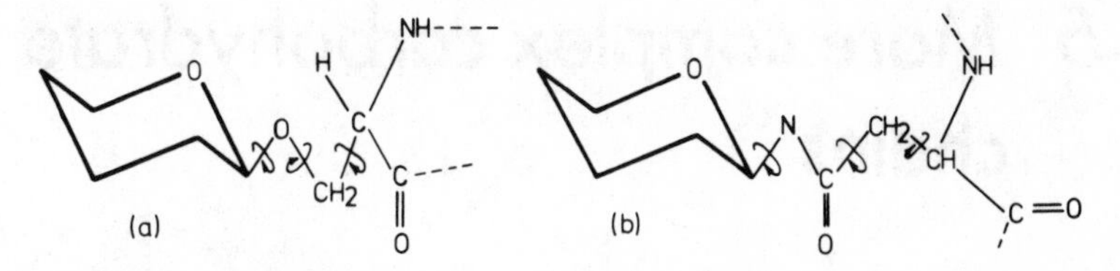

Fig. 4.15 Common linkages of carbohydrate chains to polypeptide in proteoglycans and glycoproteins; (a) *O*-glycosidic to serine; (b) *N*-glycosidic to asparagine.

similar to the 1,6-interglycosidic type in that they also have a separation of three bonds about which rotation is possible (see Fig. 4.15).

Loosely-jointed linkages also occur in linear or essentially linear chains but these are not common. When they occur the flexibility of these chains must represent a drive to favour the disordered rather than the ordered conformation in solution (Section 3.3). As we have seen, the disordered form is possible and sometimes even preferred for other types of carbohydrate chains as well, whatever the linkages – and this may be so under biological as well as non-biological conditions. Random-coil chains do in fact have important biological functions which depend on their typical properties such as the ability to bind water and ions and to interact with other polymer chains [27].

A good example of water binding by carbohydrate chains is in the connective tissues of animals which owe their form and properties to a meshwork of collagen fibres within which are trapped complex molecules (proteoglycans – see Chapter 5) which carry polyanionic carbohydrate chains which seem to exist substantially in the random coil form. Exactly the same driving force (conformational entropy) that favours the random coil (Section 3.3), must also favour a large aqueous domain for each chain to allow the shape to fluctuate. This is reinforced by the influence of mobile cations retained in the neighbourhood of the polyelectrolyte chain by the need for overall electrical neutrality, which are similarly driven by entropy to mix to the maximum extent with solvent. All this generates an internal osmotic pressure by which water is imbibed to create turgor which acts on the collagen meshwork to keep it swollen, as required for the physiological functioning [32]. For example, this imparts the characteristic tissue flexibility (e.g. of cartilage) and maintains an open structure to allow traffic in large and small molecules.

It is instructive to contrast the function of these carbohydrate chains in absorbing and retaining about 25 parts by weight of water, with that of the starch granule which is essentially dehydrated. The conformational entropy is offset in the granule by cooperative interactions which keep the chains closely packed and therefore the overall structure is relatively dehydrated. What would otherwise happen is easily seen when starch granules are gelatinized by heating in water to 'melt' the ordered state; the granule expands enormously in volume as large amounts of water are sucked in.

5 More complex carbohydrate chains

We now move on to consider carbohydrate sequences with more than one type of sugar unit and/or more than one type of linkage. Their properties can be correlated with whether the particular sequence is of the periodic, interrupted, or aperiodic type, as defined in Section 3.1 and illustrated in Chart 2 (pp. 32–33).

5.1 Periodic chains with mixed linkages

The simplest examples of chains in which two or more types of unit are present in a regular periodic arrangement are the alternating sequences (Fig. 5.1) in algal polysaccharides such as agarose and carrageenans, and polysaccharide chains of animal connective tissues such as hyaluronate and chondroitin 6-sulphate. Yet more complex periodic sequences are common in the cell surface structures of bacteria, for example in the antigenic lipopolysaccharides of gram-negative bacteria such as *Salmonella*, where up to five or more sugar units may be present in the repeating units (Chart 2, p. 32–33).

The polymers shown (Fig. 5.1) are members of a large family [27] in which units are alternately linked to position 3 and position 4 and all bonds between units are equatorial. Actually, the sequences are often complicated to some extent by deviations from a strictly regular pattern, as we shall discuss in Section 5.2.4. With the possible exception of hyaluronate, the carbohydrate chains of this type which occur in animals are attached to a polypeptide backbone [33] by linkages discussed earlier (Section 4.4) in complex macromolecules known as proteoglycans. The sugar units in these alternating polymers belong to different conformational families: the 4-linked unit if present alone in a homopolysacharide would generate a ribbon whereas the 3-linked unit would generate a hollow helix. Thus the general nature of any ordered shape is not easily predicted by mere inspection of the formula. Computer-model building calculations show that possibilities actually range from undulating ribbons to extended hollow helices [34]. It turns out that the shapes characterized in the condensed phase by X-ray diffraction are all variants of an extended hollow helix. These differ in degree of extension (Fig. 5.2) and this influences the packing in the same way as it does for simpler hollow helices (Fig. 4.12). The less extended members are agarose and carrageenan, which pack in double helices with parallel chains. These double helices are different in detailed structure: agarose is more compressed and, having the 4-linked unit in the L-form rather than the D-form as in carrageenan, the chain sense is left handed rather than right handed. Other members are sufficiently extended to

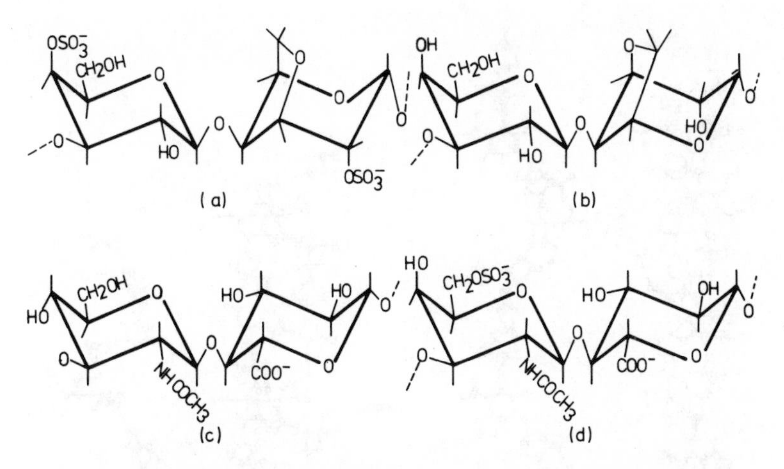

Fig. 5.1 Formulae of some alternating polysaccharides: (a) ι-carrageenan, polymer of 3-linked β-D-galactopyranose 4-sulphate and 4-linked 3,6-anhydro-α-D-galactopyranose 2-sulphate; (b) agarose, polymer of 3-linked β-D-galactopyranose and 4-linked 3,6-anhydro α-L-galactopyranose; (c) hyaluronate, polymer of 3-linked 2-acetamido-2-deoxy-β-D-glucopyranose and 4-linked β-D-glucopyranosyluronate; (d) chondroitin 6-sulphate, polymer of 3-linked 2-acetamido-2-deoxy β-D-galactopyranose 6-sulphate and 4-linked β-D-glucopyranosyluronate.

pack by nesting. This distinction has important implications for the biological state and properties: agarose and carrageenan occur in different species of marine red algae as thick extracellular gels to maintain the osmotic environment, give physical protection, and permit transport of metabolites to and from the cells. There is now much evidence that these gel structure are formed by the cross-linking of chains by double helices identical or at least closely similar to those that exist in the solid state [35]. These double helices can be very stable under aqueous conditions and thus it is not surprising that the gels are permanent structures which have an elastic response to deformation i.e. their internal structures do not readily rearrange.

Hyaluronate and chondroitin 6-sulphate proteoglycan are extracellular in animals, especially as components of connective tissues (such as cartilage and the lower layer of skin) in which large amounts of water are bound. As with agarose and carrageenans, water binding in the biological state is matched by an ability when isolated to support large amounts of water in an open, porous gel or gel-like structure. The important mechanisms of water binding by the animal carbohydrate chains, depend on chain flexibility as explained in Section 4.4. If specific associations exist between chains in these gel structures, the rheological properties show that they can only have a fleeting existence. For example, the hyaluronate gel is viscoelastic rather than purely elastic, which means that it flows slowly under stress. This points to a dynamic 'making and breaking' character in any chain associations. Viscoelastic properties are

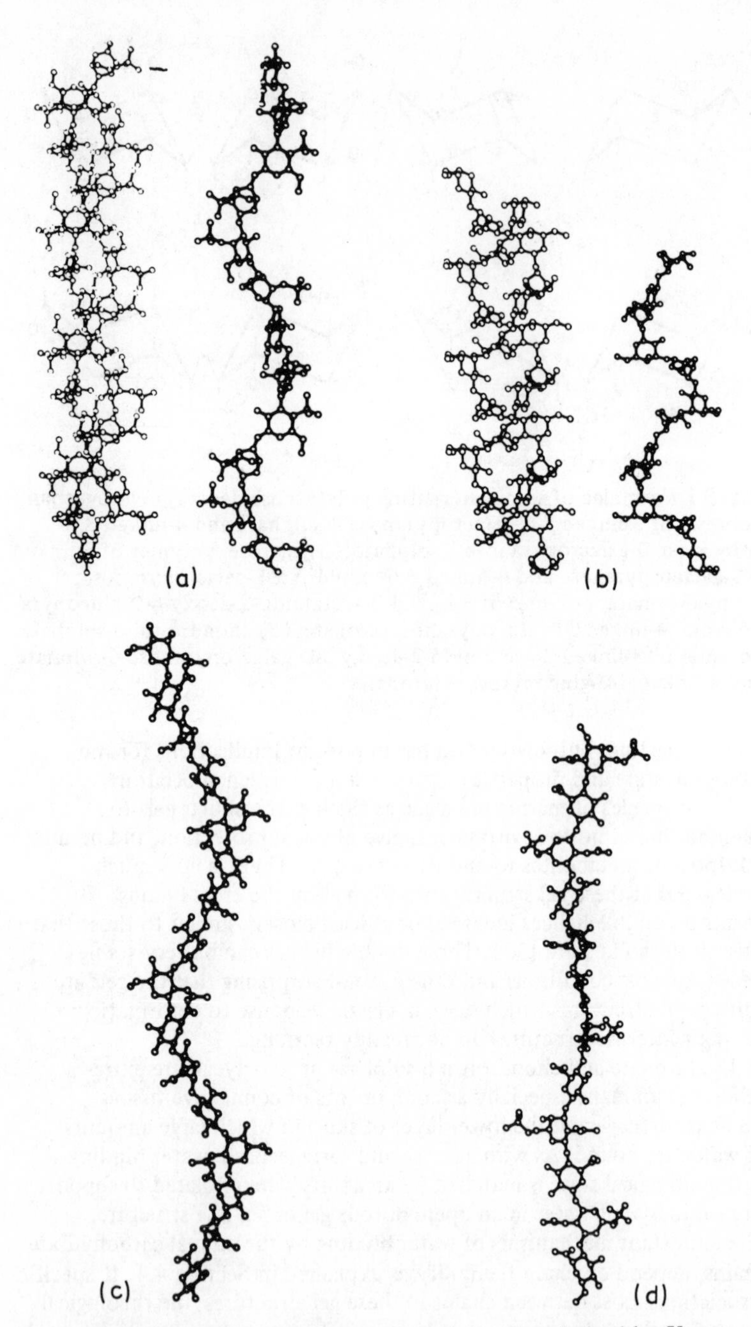

Fig. 5.2 Helices of some alternating polysaccharides, derived by X-ray fibre diffraction analysis. Each double helix structure is shown with a component strand. (a), (b), (c), (d) correspond to Fig. 5.1. All drawn to the same scale.

appropriate to the biological function because the viscous component would permit the relative movement of organs or parts of them; for example, joint surfaces could move in locomotion. At the same time, the elastic component would provide support as is also necessary, and even protection against damage by sudden movement or shock [36].

Connective tissues of different ages and from different organs contain different amounts of hyaluronate, and of chondroitin 6-sulphate and related proteoglycans. All these carbohydrate chains are based upon the same type of alternating sequence (see also Section 5.2.4) and, when crystallized for investigation by X-ray diffraction, are found as variants of a similar type of extended left-handed hollow helix. [37] Each type of chain can exist in different forms of this helix, depending on the conditions. The biological importance of ordered shapes is not clear although there is evidence for specific interactions [33] with protein components of tissues and between the chains themselves, in which it seems likely that they could be involved. The making and breaking character of the carbohydrate associations would certainly be consistent with the 'nested' arrangement found in the solid state rather than the double-helix associations in agarose and carrageenan gels, because the breaking of nested associations would not have the topological problems inherent in the separation of strands from a double helix.

The analogy is striking with the plant cell wall in which another composite structure is built up from a number of variants on a common shape (in that case, a ribbon), and biological variation is provided by mixing components in different proportions. In connective tissues, we know that the properties of chains are important in their disordered states (Section 4.4), in addition to any significance that may exist for ordered forms.

5.2 Interrupted chain sequences

5.2.1 Carrageenans

The next carbohydrate chains in order of stereochemical complexity are those in which sequences with repeating regularity are separated by regions having a different type of regularity or no regularity at all. Many of the sequences discussed earlier in this book are actually segments of regular structure within longer chains of just this type, rather than distinct polymers themselves.

A good illustration of the influence of interruptions is provided by carrageenan, in which the alternating sequence (Fig. 5.1) is interrupted from time to time by galactose 2,6-sulphate residues replacing a proportion of 3,6-anhydrogalactose 2-sulphate. This has profound implications for the overall shape, as would be expected from Chart 1, (pp. 28–29) which shows that the replacement is of a unit with the $^1C_4$ conformation linked by two equatorial bonds, by one which is $^4C_1$ and linked by two axial bonds. Computer-model building confirms that galactose sulphate must introduce a kink into the regular helical sequence. The carrageenan chain is then pictured as having regions of regular

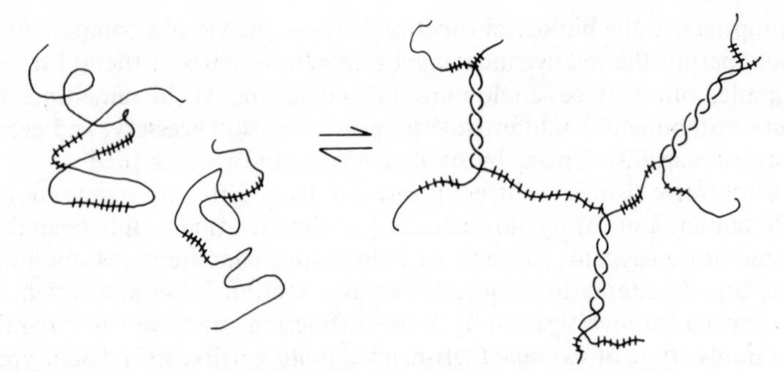

Fig. 5.3 Schematic picture of coil to double helix conversion with gel formation. Helix-forming sequences are represented by the continuous parts of each line and kinking sequences by the crossed parts.

sequence with the ability, under favourable conditions, to enter into double helix formation, separated by other sequences which act as kinks. A typical chain may contain 8 or 10 helix-forming regions. In the conversion to helix in solution, each of these must nucleate independently and need not involve the same chain partner. The result is then a cross-linked network structure (Fig. 5.3) and this is why the solution sets to a gel – an infinite three dimensional framework exists to resist liquid flow. Carrageenan preparations can be made without kinks, for which the coil to double-helix conversion does not cause gelation – no doubt because the entire length of each chain can pass into a continuous helix and molecules in solution then exist as independent pairs. When the kinks force the chains into a cross-linked network, they therefore represent a biological device which is necessary for gel properties to develop. Other properties determined by the kinks include the size and shape distribution of pores and channels which influence transport through the gel, and the rigidity and hence quality of mechanical protection to cells embedded within it. These properties seem to be under biological regulation, through control of the proportion and distribution of kink units.

In the biosynthesis of carrageenans, 3,6-anhydride units are introduced at a late stage into chains that are already polymerized and sulphated. This means that the chain is first synthesized in a form that is not able to convert to double helix, and helix forming regions are then generated by conversion of galactose 2,6-disulphate units of 3,6-anhydrogalactose 2-sulphate. This is analogous to the biosynthesis of collagen which is known to be through a non-helical and therefore, soluble and easily transported precursor, which is modified near the site of deposition to convert to helical form. The reaction by which anhydride units are introduced into carrageenan is shown in Fig. 5.4; the driving force to close the anhydride ring is derived from the splitting of the sulphate ester. We now see that the kinks in carrageenan are actually relics which remain from the precursor polysaccharide. For obvious reasons, the enzyme which catalyses the reaction in Fig. 5.4 is trivially named 'de-kinkase' [38].

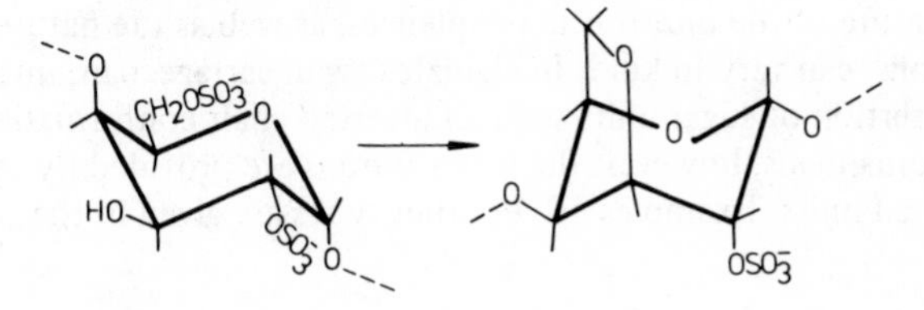

Fig. 5.4 Conversion of galactose 2,6-disulphate unit to 3,6-anhydrogalactose.

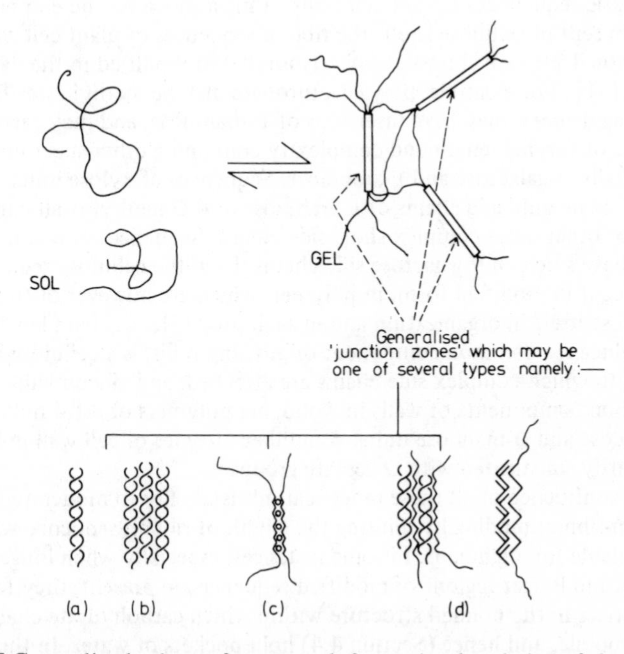

Fig. 5.5 Generalized scheme for network formation by chains of the interrupted type: (a) cross-linkage by double helices as in ι-carageenan; (b) by bundles of double helices as in agarose; (c) by ribbon–ribbon associations of the egg-box type as in alginate; (d) helix–ribbon associations as in mixed systems containing agarose and certain galactomannans, or xanthan (a bacterial polysaccharide) and certain glucomannans.

5.2.2 Interrupted sequences in the general mechanism of gelation [35]

Carrageenan can be regarded as an archetype for a large family of carbohydrate chains which form gel structures because of the interrupted regularity in their covalent sequence. From the classification of conformational families of carbohydrate chains in Chapter 4, we could expect that associations could be of a number of different types, involving helix-helix, ribbon-ribbon or helix-ribbon associations. Some examples which have been well characterized are shown in Fig. 5.5.

The nature of the interrupting sequences, as well as the nature of the associations, can vary in kind. In alginates, as in carrageenan, interruptions are by insertion of sugar units with an inverted chair conformation. In galactomannans, however, the interruptions are provided by sequences of branched units. Examples of yet other types are given in the following section.

5.2.3 Interrupted sequences in the plant cell wall

It is beyond the scope of this small book to describe all the variations on the basic sequences of plant cell walls. It must, however, be emphasized that except in cellulose itself, the ribbon sequences of plant cell walls (Section 4.2.3) are almost always elaborated or modified in the natural state [31]. For example, polygalacturonate may be methyl-esterified to varying degrees, may have insertions of L-rhamnose, and may carry side chains of varying length and complexity containing other sugar units especially D-galactose and L-arabinose. Sequences of xylose units usually carry single unit side chains of L-arabinose or 4-*O*-methyl-D-glucuronic acid or other more complex short side chains. Sequences of mannose may have single unit galactose side chains. Even the cellulose sequence can occur in modified form, in polymers which are however distinct from cellulose itself in organization and in such properties as chain length; xyloglucan, a common component of growing walls, is a cellulose-like chain to which complex side chains are attached, and glucomannans, common components of walls in wood, are polymers of β-1,4 linked D-glucose and D-mannose units. A number of types of cell wall polymer are partly substituted with *O*-acetate groups.

The influence of all these modifications is usually to moderate the ribbon-ribbon binding by limiting the length of ribbon sequence which is available for alignment. In some instances, especially when longer side chains and longer regions of modified sequence are present, they form interstices in the bonded structure within which carbohydrate chains are free, mobile, and hence (Section 4.4) hold pockets of water. In this way, a cell wall structure is built up which is not merely a solid case like a concrete or cast-iron shell but combines the cohesion derived from binding with other properties such as porosity, hydration and some flexibility.

5.2.4 Interrupted sequences in carbohydrate chains of connective tissues [33]

Just as in plant cell walls the sequences are rarely found in entirely regular form, so in connective tissues the chains usually deviate from the idealized alternating arrangement of their helical prototypes. There is apparently one exception with a totally periodic sequence in each of the two systems – cellulose in plant cell walls and hyaluronate in the animal tissues. For example, chondroitin 6-sulphate (Fig. 5.1(d)) may occur with the 3-linked unit partly 2,6-disulphated and/or nonsulphated, and with the 4-linked unit partly sulphated. Another polymer of this family, dermatan sulphate is often represented as an alternating chain of 3-linked

2-acetamido-2-deoxy-β-D-galactopyranose 4-sulphate and 4-linked α-L-idopyranosyluronate units (see Chart 1 for sugar formulae) but in fact the former unit may occur in the various states of sulphation and the latter may in part be sulphated or even replaced by 4-linked β-D-glucopyranosyluronate. Likewise keratan sulphate is an alternating chain of 4-linked 2-acetamido-2-deoxy-β-D-glucopyranose and 3-linked β-D-galactopyranose, with both units partly sulphated on position 6. The final member of this family is chondroitin 4-sulphate, for which the idealized representation is the same as chondroitin 6-sulphate (Fig. 5.1) except that the sulphate is on position 4 of the 3-linked unit rather than position 6. For all these chains, there is a special sequence of units in the so-called 'linkage region' by which the chain is attached to the polypeptide. Except for keratan sulphate, this is galactose-galactose-xylose-serine with all three sugars being 3-linked and β.

The characterization of conformations by X-ray diffraction has so far been done mainly with chains having sequences close to the idealized arrangements. (These can be obtained by careful selection of the biological source, followed by fractionation.) The chains are found to have more conformational versatility than the plant wall polymers – for example, one member can sometimes be induced to mimic the conformation of another. Perhaps the variations here have a more subtle function than usual – such as, to provide mechanisms not only for the making and breaking of ordered shapes and association but also for modification and fine control of the geometry and strength of interaction.

5.3 Aperiodic sequences

This large and important class of carbohydrate chain is probably represented in all the cells of every organism, as well as in most extracellular fluids in the animal kingdom. Almost always these chains are attached to polypeptide and/or lipid. Despite this wide occurrence and the crucial biological functions with which they are associated such as immune systems, receptors on cell surfaces for hormones and neurotransmitters, and recognition molecules involved in the processes of cell sorting during tissue development, little definite information is available either on the shapes of these carbohydrate chains or the biological part they play at the molecular level.

The carbohydrate chains of most animal glycoproteins can be grouped in two main categories [39]:–

(i) *O*-Glycosidic type: these are made up mostly of D-galactose, 2-acetamido-2-deoxygalactose and/or neuraminic acid units. The attachment to protein is from 2-acetamido-2- deoxygalactose to serine or threonine (see Fig. 4.15). When neuraminic acid occurs, it is at the outermost terminal(s). These chains are often, but not always short (four sugar units or fewer).

(ii) *N*-Glycosidic type: these contain a chain made up from D-mannose and 2-acetamido-2-deoxyglucose units linked to protein through a linkage from the latter sugar to asparagine (see Fig. 4.15). To this 'core' chain may be added further chains with galactose, neuraminic acid and/or fucose units. An example is the carbohydrate chain of the immunoglobulin IgG, shown in Chart 2 (pp. 32–33).

Aperiodic structures of other types also occur naturally, sometimes with different linkages to protein or lipid, but these two categories include a very large number (perhaps the majority) of natural structures. Our discussion of structure and function will be limited to these.

5.3.1 Mucous secretion

In vertebrates, the cavities which communicate with the environment – such as the digestive and respiratory tracts and the uterus – are covered by a form of mucus which contains glycoproteins. This is believed to protect against chemical, physical and microbiological injury and also to lubricate for movement. One example is the layer of mucous gel which lines the stomach to protect the mucosa from acidic stomach contents [40]. This gel is formed from a glycoprotein of very high (2 million) molecular weight, which contains a very high proportion of carbohydrate (74%) relative to protein. The carbohydrate chains are mostly or entirely of the *O*-glycosidic type to serine and threonine, but with rather longer (10–15 sugar units) and more complicated sequences than usual. Although based on galactose and 2-acetamido-2-deoxygalactose they also contain some sulphate ester, fucose and 2-acetamido-2-deoxy-glucose units.

Physical measurements on dilute solutions of this glycoprotein in aqueous salt, show that its 'effective size' is very much larger than expected for a dense globular particle of this molecular weight; in fact the size is such that the solution volume should be completely filled at a concentration of about 25 milligrammes of glycoprotein per millilitre of solution. At precisely this concentration range, the solution begins to show gel-like properties. From this evidence it follows that the glyco-protein is an expanded bushlike molecule which entrains a large volume of solvent within its molecular outline, and can 'stick when it touches' to other like molecules, to link up to form the gel network. It is found that the effective molecular size can be altered by changing the salt concen-tration, and this leads us to suspect some degree of flexibility (i.e. dis-order) in the conformation. The actual mechanism which causes the molecules to join is quite unknown. It is not even clear whether it involves carbohydrate-carbohydrate, protein-protein, or carbohydrate-protein contacts.

From the glycoprotein concentration in the gel, it is likely that the network would be sufficiently dense to limit penetration through it of proteolytic enzymes from the digestive fluids to the mucosa, and also to reduce turbulence so that a relatively stable pH gradient can be main-tained and therefore the pH at the mucosa need not be so low as in the digestive fluid. In both of these ways, the glycoprotein gel is believed to protect the mucosa from harmful attack or irritation.

5.3.2 Carbohydrates on folded polypeptide chains: immunoglobulin IgG

Although carbohydrate chains frequently occur in attachment to proteins

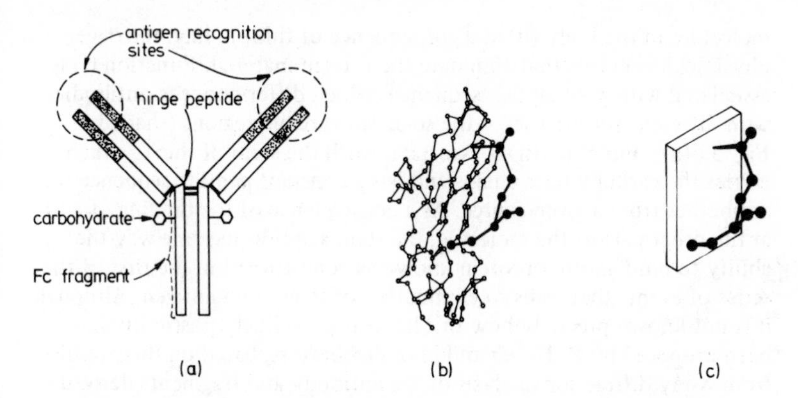

Fig. 5.6 Representation of antibody (IgG) molecule: (a) schematic to show relative locations of the different parts; (b) contour of the carbohydrate chain (heavy lines – circles represent centres of each sugar ring) and neighbouring regions of peptide (thinner lines – circles represent C_α fragment) [43]; (c) schematic of (b), showing how the carbohydrate chain projects from the peptide sheet (shown as a parallelepiped) to cover one face.

with important functions as enzymes, hormones and antibodies, there is little evidence that the carbohydrate is ever part of the 'active site'. It does seem possible however, that the carbohydrate may act as a 'conformational modifier' of the protein. For example, certain enzymes (e.g. yeast invertase, β-*N*-acetylhexosaminidase, α-galactosidase) remain active when their carbohydrate chains are split, but become much more prone to denaturation suggesting that a source of stability has been removed with the carbohydrate [41]. As evidence has accumulated on peptide sequences around the attachment sites of carbohydrates, it has become quite strikingly clear that they are often of a particular sort that, by comparison with known protein conformations, would be predicted to adopt the peptide conformation known as the β-turn [42]. This fragment of protein structure resembles a hairpin bend, and allows the peptide chains on either side to align – for example as a pair of interacting α-helices or to form a β-pleated sheet. However, until recently, no three-dimensional structure had been characterized for a glycoprotein and therefore it was impossible to examine the relation between protein and carbohydrate shapes in a rigorous way.

The crystal structure has now been determined for a glycoprotein with an *N*-glycosidically linked carbohydrate chain which does indeed turn out to be attached at a β-turn and to have an intimate relationship with the protein part. This structure is a carbohydrate-bearing fragment of the antibody molecule IgG [43]. The antibody itself is a very large glycoprotein (molecular weight about 170 000) with one type of carbohydrate chain which has a sequence [39] of the type shown in the last entry of Chart 2 (pp. 32–33). The molecule is roughly Y-shaped and contains two conformationally equivalent halves (Fig. 5.6(a)). Its biological functions are (i) to recognize and bind foreign cells and macromolecules in the body (ii) as a consequence of this binding, to trigger physiological events that eliminate the foreign material. Function (i) is associated with polypeptide sequences which differ between antibodies with different specificities – the so-called variable regions (shaded in Fig. 5.6(a)). Function (ii) is associated with the stem of the Y, which carries the carbohydrate chains and has a constant peptide sequence for antibodies from a given source. As a consequence of the binding of antigen at the other part of the molecule, the stem acquires in some way the ability to bind another protein known as complement in the first of the series of events that leads to elimination of the bound antigen. Although it is not known precisely how this happens, one likely possibility has been proposed by R. Huber and his collaborators, based on their results from X-ray diffraction analysis of the antibody and fragments derived from it [43].

The so-called F_c fragment (Fig. 5.6(a)), which is prepared by splitting the antibody with the enzyme pepsin, crystallizes with both the polypeptide and the carbohydrate chains in ordered conformations. At time of writing, the analysis is not yet at a stage at which the conformation and interactions of the carbohydrate chain can be defined in detail but the outline is clear enough (Fig. 5.6(b)). The carbohydrate is attached to a bend in the polypeptide chain which is arranged in a sheet structure. The ordered carbohydrate chain makes intimate contact with a number of amino acid units, suggesting mutual stabilization of conformations (although the precise interactions such as possible hydrogen bonds, cannot be defined yet), and it covers a large part of the surface of the polypeptide sheet (Fig. 5.6(c)). When the antibody itself crystallizes however, the F_c part is conformationally disordered from the hinge region down. Huber therefore proposes that the ordered form seen in crystalline F_c fragments is characteristic of the antibody bound to antigen – rather than of the unbound antibody. This would be consistent with certain other physical evidence for a 'stiffening' of antibody when it binds antigen in solution, and would also explain how it is that complement binds to the antibody only when antibody has itself bound antigen: presumably, complement can only bind when the F_c region has become conformationally ordered.

No diffraction evidence is available yet for an antigen-antibody complex, but it is possible to build a model of the antibody in the bound state by putting together the known structures of the F_c and other fragments based on the assumption of the 'stiffening' mechanism that we have just described. When this is done, the carbohydrate chain acquires even more contacts with peptide – especially with units of the hinge region – and it is seen that the antibody molecule converts from a Y to a T shape [43].

In summary, the binding of antigen appears to convert the antibody from a 'flexible Y' to a 'rigid T' and the carbohydrate chain helps to 'key in' the cross member and the upright to stabilize the T. Over the next few years, it will be interesting to see to what extent it is general for carbohydrate chains in glycoproteins to be situated on bends in the protein structure, and whether they also adopt ordered conformations which screen the protein surface and influence conformation changes with a trigger function in biological response.

5.3.3 Glycoproteins as antigenic determinants and 'recognition molecules' [44]

This discussion of glycoproteins would be incomplete without mention of the well known importance of their carbohydrate chains as 'recognition marks'. For example, blood group activity is an expression of the terminal sequence of the carbohydrate chains of glycoproteins and glycolipids on the surfaces of red blood cells. Again, it seems that the absorption into the liver of plasma glycoproteins and of red blood cells with subsequent breakdown, is prompted by removal of a terminal neuraminic acid unit to expose an interior sugar which is the signal for uptake. In this way the carbohydrate chains are involved in the control of turnover rate.

These properties seem to depend on the local carbohydrate sequence over a few units, rather than the overall shape of the biopolymer in three dimensions.

References

[1] Oosterhoff, L. J. (1971), *Pure Appl. Chem.*, **25**, 563–571.
[2] *For further details and references to original literature, see Stoddart, J. F. (1971), *Stereochemistry of Carbohydrates*, John Wiley, New York.
[3] Arnott, S. (1970), *Progr. Biophys. Mol. Biol.*, **6**, 265–319 Arnott, S. and Hukins, D. W. L. *Biochem. J.*, **130**, 453–465.
[4] *Ferrier, R. J. and Collins, P. M. (1972), *Monosaccharide Chemistry*, Penguin Books, Harmondsworth, Middlesex.
[5] Phillips, D. C. (1966), *Scientific American*, **215**, 78–90.
[6] Angyal, S. J. (1969), *Angew. Chem. Internat. Ed.*, 8, 157–166.
Rees, D. A. and Smith, P. J. C. (1975), *J.C.S. Perkin II*, 830–835; Dunfield, L. G. and Whittington, S., *ibid.*, in press.
[7] *Rees, D. A. (1973), *MTP Internat. Rev. Sci., Org. Chem., Series One*, 7, 251–283.
[8] Rees, D. A. and Smith, P. J. C. (1975), *J. C. S. Perkin II*, 836–840.
[9] Rees, D. A. and Thom, D. (1977), *J.C.S. Perkin II*, 191–201.
[10] St.-Jacques, M., Sundararajan, P. R., Taylor, K. J. and Marchessault, R. H. (1976), *J. Amer. Chem. Soc.*, **98**, 4386–4391.
[11] Rees, D. A. and Scott, W. E. (1971), *J. Chem. Soc.* (B), 469–479.
[12] Esau, K. (1965), *Plant Anatomy*, (second edition), Wiley, New York.
[13] Palevitz, B. A. and Hepler, P. K. (1976), *Planta*, **132**, 71–93.
[14] *For a stimulating and speculative discussion of polysaccharides in the growing plant cell wall, see Albersheim, P. (1975), *Scientific American*, 80–95, Scientific American Inc., New York.
[15] Gardner, K. H. and Blackwell, J. (1974), *Biopolymers*, **13**, 1975–2001.
[16] Kolpak, F. J. and Blackwell, J. (1976), *Macromolecules*, **9**, 273–278.
[17] Settineri, W. J. and Marchessault, R. H. (1965), *J. Polymer Sci., Part C*, **11**, 253–264.
[18] Atkins, E. D. T., Nieduszynki, I. A., Mackie, W., Parker, K. D. and Smolko, E. E. (1973), *Biopolymers*, **12**, 1879–1887.
[19] Rees, D. A. and Wight, A. W. (1971), *J. Chem. Soc. (B)*, 1366–1372.
[20] Kohn, R. (1975), *Pure Appl. Chem.*, **42**, 371–397; Grant, G. T., Morris, E. R., Rees, D. A., Smith, P. J. C. and Thom, D. (1973), *FEBS Letters*, **32**, 195–198.
[21] Rudall, K. M. (1969), *J. Polymer, Sci., Part C*, **28**, 83–102.
[22] Oldmixon, E. H., Glauser, S. and Higgins, M. L. (1974), *Biopolymers*, **13**, 2037–2060.
[23] Kainuma, K. and French, D. (1972), *Biopolymers*, **11**, 2241–2250; Wu, H.-C. H. and Sarko, A. (1977), *Carbohyd. Res.*, in press.
[24] Murphy, V. G., Zaslow, B. and French, A. D. (1975), *Biopolymers*, **14**, 1487–1501.
[25] French, A. D. and Murphy, V. G. (1977), *Polymer*, in press.
[26] Bergeron, A., Machida, Y. and Bloch, D. (1975), *J. biol. Chem.*, **250**, 1223–1230; and earlier papers cited there.
[27] *Rees, D. A. (1975), *MTP Internat. Rev. Sci., Biochem., Series One*, **5**, 1–42.
[28] Smith, W. L. and Ballou, C. E. (1973), *J. biol. Chem.*, **248**, 7118–7125; and earlier papers cited there.
[29] Bluhm, T. L. and Sarko, A. (1976), *Canad. J. Chem.*, in press; Marchessault, R. H., Deslandes, Y., Ogawa, K. and Sundararajan, P. R. (1976), *Canad. J. Chem.*, in press.

[30] Atkins, E. D. T. and Parker, K. D. (1969), *J. Polymer. Sci., Part C*, **28**, 69–81.

[31] *Aspinall, G. O. (1970), *Polysaccharides*, Pergamon, Oxford.

[32] Wells, J. D. (1973), *Proc. Roy. Soc. Series B*, **183**, 399–419.

[33] *Muir, H. and Hardingham, T. E. (1975), *MRP Internat. Rev. Sci., Biochem., Series One*, **5**, 153–222.

[34] Rees, D. A. (1969), *J. Chem. Soc. (B)*, 217–226.

[35] *Rees, D. A. and Welsh, E. J. (1977), *Angew. Chem. Internat. Ed.*, in press.

[36] Balazs, E. A. and Gibbs, D. A. (1970), in *Chemistry and Molecular Biology of the Intercellular Matrix* (Ed.) E. A. Balazs, Volume 3, 1241–1253, Academic Press, New York.

[37] Winter, W. T., Guss, J. M., Hukins, D. W. L. and Arnott, S. (1976), *J. Mol. Biol.*, in press, and earlier papers cited there; Sheehan, J. K., Atkins, E. D. T. and Nieduszynki, I. A. (1975), *J. Mol. Biol.*, **91**, 153–163 and earlier papers cited there.

[38] Lawson, C. J. and Rees, D. A. (1970), *Nature* (London), **227**, 392–393.

[39] *Kornfeld, R. and Kornfeld, S. (1976), *A. Rev. Biochem.*, **45**, 217–237.

[40] Allen, A., Pain, R. H. and Robson, T. R. (1976), *Nature* (London), **264**, 88–89 and earlier papers cited there.

[41] *Gottschalk, A. (1972), in *Glycoproteins* (second edition, (Ed.) A. Gottschalk), **Part B**, 1294–1298.

[42] Aubert, J.-P., Biserte, G. and Loucheux-Lefebre, M. H. (1976), *Arch. Biochem. Biophys.*, **175**, 410–418.

[43] Huber, R., Deisenhofer, J., Colman, P. M., Matsushima, M., and Palm, W. (1976), *Nature* (London), **264**, 415–420.

[44] *Hughes, R. C. (1976), *Membrane Glycoproteins: A Review of Structure and Function*, Butterworths, London.

**Books and articles for further reading*

The references marked with an asterisk in the above list give more detailed discussion of particular topics in a style that is suitable for further study. The others provide documentation and evidence for reference and are less suitable for general reading.

Index